Using Federalism
to Improve
Environmental Policy

T0273245

AEI STUDIES IN REGULATION AND FEDERALISM
Christopher C. DeMuth and Jonathan R. Macey, series editors

Using Federalism to Improve Environmental Policy

Henry N. Butler and
Jonathan R. Macey

The AEI Press

Publisher for the American Enterprise Institute
WASHINGTON, D.C.

1996

Library of Congress Cataloging-in-Publication Data

Butler, Henry N.
    Using federalism to improve environmental policy / Henry N. Butler
and Jonathan R. Macey.
        p.   cm. — (AEI studies in regulation and federalism)
    Includes bibliographical references.
    ISBN 0-8447-3962-6 (c   — ISBN 0-8447-3963-4 (pbk. :

    1. Environmental law—United States.   2. Federal government—
United States.   3. Decentralization in government—United States.
I. Macey, Jonathan R.   II. Title.   III. Series.
KF3775.B88   1996
344.73'046—dc20
[347.30446]   96-16419
                                                            CIP

1   3   5   7   9   10   8   6   4   2

THE AEI PRESS
Publisher for the American Enterprise Institute
1150 17th Street, N.W., Washington, D.C. 20036

Contents

1
Introduction

More than twenty years of intensive federal regulation of environmental risks have demonstrated the severe drawbacks of centralized environmental policy. The command-and-control regulatory strategy that dominates environmental policy has proved inadequate: it has not set intelligent priorities, it has squandered resources devoted to environmental quality, it has discouraged environmentally superior technologies, and it has imposed unnecessary penalties on innovation and investment.[1]

There is no shortage of policy proposals for improving environmental quality by changing these federal strategies.[2] Indeed, for many issues there is a surprising consensus about how to change current policies. Thus, the real question is why the federal government has been so slow to respond to both the widely identified problems and the widely recognized solutions. Part of the answer is that the high degree of centralization in federal environmental regulation has led to inflexibility and inertia.[3]

This study considers whether environmental policy can be improved by changing the allocation of authority for environmental regulation within our federal system. Decentralization through greater reliance on market incentives and economic property rights, through

---

1. See generally Bruce A. Ackerman and Richard B. Stewart, "Reforming Environmental Law," *Stanford Law Review* 37 (1985): 1,333.

2. See Ralph A. Luken, *Efficiency in Environmental Regulation: A Benefit-Cost Analysis of Alternative Approaches* (Boston: Kluwer, 1990).

3. Another explanation, however, is the self-interest of the regulators and the regulated alike in maintaining the status quo. The possible administrative and political routes past these obstacles to reform are beyond the scope of this volume. Our purpose is to demonstrate that it is indeed a legitimate goal to motivate such reform.

greater state control over environmental policy, or through both appears to hold promise as a source for more flexible, dynamic, and responsive environmental policy.[4] Although a few specific proposals for reforming environmental law will be discussed in detail, our primary focus is on the institutional setting in which environmental policies are adopted. Specifically, under what circumstances can environmental policy be improved—or made more efficient—by reallocating regulatory authority among the various state and federal actors?

Our discussion proceeds as follows. In chapter 2, we review the evolution of federal domination of environmental regulation. In chapter 3, we summarize and critique the leading rationales for federal environmental regulation. In chapter 4, we consider the incentives of politicians at different levels of government to meet the environmental quality preferences of their constituents. We develop principles for the most efficient governmental level to regulate different types of environmental concerns. Our analysis suggests that generally the size of the geographic area affected by a specific pollution source should determine the appropriate governmental level for responding to the pollution. The regulating jurisdiction need not be larger than the regulated activity. In other words, when a polluting activity is limited to a particular locality or state, there is little justification for federal environmental regulation. When a federal government response is justified, it should be the most limited response possible.

In chapter 5, we use our model of environmental regulatory federalism to assess and recommend improvements for the existing pattern of environmental regulation in air pollution, water pollution, and land pollution. We argue that, in every area of pollution, environmental regulation has been centralized beyond any possible justification resulting in tremendous costs.

In chapter 6, we summarize our analysis and offer several recommendations for change in the allocation of regulatory authority. There is a fairly general consensus that the current allocation of regulatory authority is not getting the job done in a satisfactory man-

---

4. See Ackerman and Stewart, "Reforming Environmental Law." See also Terry L. Anderson and Donald R. Leal, *Free Market Environmentalism* (Pacific Research Institute, 1991), and David Young, "Expanding State Initiation and Enforcement under Superfund," *University of Chicago Law Review* 57 (1990): 985.

ner. We are confident that greater state and local control of environmental regulation would lead to a larger net benefit from environmental regulation because of the institutional incentives to find the lowest cost methods of reducing pollution and protecting the environment. Although the public is unwilling to oversee the intricacies of federal environmental regulation,[5] we believe the public would be receptive to the logic of our arguments.[6] Moreover, we think that the American public's desire for aggressive environmental enforcement can be satisfied better by radical restructuring of environmental regulatory authority.[7]

---

5. See, for example, Dwight R. Lee, "Politics, Ideology, and the Power of Public Choice," *Virginia Law Review* 74 (1988): 191, 197: "Predictably, there is little genuine public surveillance of environmental protection programs, and organized groups have significant latitude to influence (such) programs in ways that serve their private interests. This means of course that these programs are far less effective . . . than they could be." See also Everett Carll Ladd and Karlyn H. Bowman, *Attitudes toward the Environment: Twenty-Five Years after Earth Day* (AEI Press, 1995), p. 50: "The public points to the ends that public policy makers should work to achieve. The public does not think much about the means."

6. The American public has been disillusioned about the ability of the federal government to solve problems. See, for example, Ladd and Bowman, *Attitudes toward the Environment*, pp. 1–2: "Only 18 percent of Americans in a late February 1995 CBS News/*New York Times* poll say that they trust the government in Washington to do what is right just about/always/or most of the time."

7. Ibid., p. 50.

2
Federal Domination of Environmental Regulation

*Our current environmental regulatory system was an understandable re-
sponse to a perceived need for immediate controls to prevent a pollution
crisis. But the system has grown to the point where it amounts to nothing
less than a massive effort at Soviet-style central planning of the economy
to achieve environmental goals. It strangles investment and innovation. It
encourages costly and divisive litigation and delay. It unduly limits private initiative and choice. The centralized command system is simply
unacceptable as a long-term environmental protection strategy for a large
and diverse nation committed to the market and decentralized ordering.[1]*

Although the extensive command-and-control environmental regulation that emerged in the 1970s came after numerous other efforts
to control or internalize pollution externalities, the centralization of
environmental policy-making was more the result of political urgency
and frustration than of careful balancing of the costs and benefits of
centralization. Moreover, even the current centralized response to
environmental problems requires the cooperation of the states, because the centralized federal government is incapable of managing
the tremendous number of local problems encountered in regulating
local pollution. Ideally, one of the salutary characteristics of federalism would be to allow state governments to protect local environmental interests and to tailor solutions to local concerns, while the federal
government would set national standards, provide funding and expertise, and address multistate problems.

Our analysis suggests that the current problem with environmental regulation derives from an imbalance in the allocation of

---

1. Richard B. Stewart, "Controlling Environmental Risks through Economic Incentives," *Columbia Journal of Environmental Law* 13 (1988): 153, 154.

governmental functions. A survey of the evolution from the concept of a legal remedy for an environmental wrong to the proliferation of environmental statutes and regulations will reveal not only how far off balance our policies have become, but also the implications of not changing them. We begin with a brief discussion of the economics of pollution control.

The Economics of Pollution Control

The economic goal of governmental regulation of pollution is to force polluters to bear the full costs of their harsh activities. In economic jargon, the regulatory goal should be to force the internalization of externalities. Externalities are costs and benefits that are not directly priced by the market system. Since individuals in a market system respond only to the benefits and costs that they actually receive and pay for, the market system may be inadequate to deal with externalities. The market failure that results when market participants do not internalize the external costs of their activities causes resources to be misallocated. Thus, the negative externalities—or spillover costs—associated with pollution are an economic problem because they lead to an inefficient allocation of resources. Externalities in the use of resources often arise where property rights are either nonexistent or poorly specified, as is the case with the atmosphere. Such resources are free, from the perspective of the users. Those who produce products creating externalities do not pay the full cost of the resources consumed by the production of such products. Because producers will produce the quantity of goods that reflects their private costs of production, externalities lead to overproduction, which in turn leads to an inefficient overallocation of resources to the production of the good. Hence the economic justification for governmental regulation of pollution.

It is far from clear, however, that the presence of an externality is sufficient to justify governmental intervention. In reality, *externality* is a slippery concept, used more often to achieve the categorization of an event as a problem than to justify governmental intervention to solve the problem. Put another way, virtually everything that anybody does is an externality when viewed from somebody's perspec-

5

tive.[2] For example, here is an externality argument in favor of federal regulation that concerns a type of psychological externality arguably arising even when pollution does not physically cross state lines. This argument is based on the notion that all citizens of the United States may justifiably be concerned about environmental quality throughout the United States, although they are not physically exposed to local externalities in other localities. Devoted environmentalists in Oregon may be deeply concerned about the local environmental effects of chemical plants and oil refineries in Louisiana. They may argue that Louisiana's environmental laws do not adequately address the local environmental risks and therefore should be preempted by more stringent federal regulations. Thus, according to this "externality" argument, federal regulation to require more stringent local environmental regulations may be justified on the efficiency ground that purely local pollution (local in its physical damage) may actually impose additional costs in other states where citizens have stronger preferences for environmental purity. These costs can take the form of lost utility of environmentalists in states with more stringent regulations, as well as exit by polluting industries to less restrictive states. Although the existence of such interdependencies raises provocative questions about the demand for regulations and about who should bear the burden of their implementation, this argument for federal intervention is flawed for a number of reasons.

The major objection to this argument is that the local residents in Louisiana, not the Oregon environmentalists, would bear all the costs of reducing pollution. The Louisiana political process would have already indicated that the local residents believe the cost of a cleaner Louisiana to be greater than the benefits. It is tempting to assert that we are simply dealing with different sets of preferences about environmental quality, and that the national consensus is for greater environmental quality than that preferred in Louisiana. But this is not necessarily the case. In fact, every state has evinced a

---

2. When an entrepreneur creates a better manufacturing process for a particular product, rivals who are using the inferior process are harmed. From their perspective, the technological improvement is an externality. See David Haddock, Jonathan Macey, and Fred McChesney, "Property Rights in Assets and Resistance to Tender Offers," *Virginia Law Review* 73 (1987): 701, 723.

strong interest in environmental quality. Even Louisiana, often cited as an example of a state with an environmentally insensitive regulatory regime, takes great pride in its environment, as evidenced by its state motto—"the sportsman's paradise." The citizens of Louisiana would probably be happy with even higher environmental quality if they did not have to pay for it with reduced economic opportunities. Similarly, it is not surprising that the people who do not pay for higher environmental quality are in favor of more stringent standards than are the local citizens who must bear the costs. Allocation to local governments of regulatory authority over local externalities allows decisions to be made by the representatives of the citizens who benefit the most and pay the most for higher environmental quality.

Finally, the economics of pollution control demonstrates that it would be undesirable to prevent all externalities simply because many of them are the result of socially desirable economic activity.[3] Even if all negative externalities are internalized and the private cost of production equals its social cost, pollution will still not be eliminated. Instead, the result of internalization will be that those causing pollution will be required to pay the full social costs associated with their activities. Pollution is a necessary by-product of our modern lifestyle. Getting rid of waste is not free in terms of either monetary costs or the productive capacity of the nation. One of the costs of producing more man-made goods is the sacrifice of some environmental quality. Similarly, the cost of a cleaner environment is the sacrifice of some man-made goods. These observations represent the basic economic problem of scarcity—our resources are simply not sufficient to satisfy all our demands. Thus we are forced to make trade-offs. For example, only the most devout environmentalists would give up the personal freedom associated with the use of an automobile because its use causes air pollution.

The economic goal of pollution-control regulation should not be to reduce the level of pollution to zero. The goal rather should be to set the level of pollution to what it would be if producers bore all the costs created by their pollution. The combination of small externalities and of the nontrivial costs of governmental intervention suggests, however, that many externalities cannot be economically corrected.

---

3. James M. Buchanan and Craig Stubblebine, "Externality," *Economist* 29 (1962): 371.

The Common Law

The common law of torts provides several causes of action for parties harmed by the actions of others when those actions cause environmental harm. The most important action for dealing with externalities is the tort of private nuisance, which requires proof of intentional and unreasonable interference with another person's right to use of and enjoyment of land. Almost all the cases involve lawsuits between landowners located in the same geographic areas. The tort can be applied to a wide variety of externalities, including excessive noise, noxious odors, smoke or dust settling on a landowner's property, and interference with an adjoining property owner's right to sunshine to power a solar energy system. The tort of private nuisance is enforced by private lawsuits for damages and equitable relief, including injunctions.

In traditional analysis, courts balance the conflicting uses of property owners. Ownership of private property includes the right to use and enjoy one's property as long as that use does not interfere with the rights of other property owners to use and enjoy their property. Under the common law, no one has an absolute right to be free from the effects of pollution that caused no damage to property rights.

Strict liability may also be used to force private property owners to internalize the externalities they create. Although strict liability is now the majority rule for liability for injuries resulting from the sale of defective products, the theory of strict liability was originally developed to deal with injuries resulting from ultrahazardous or abnormally dangerous activities that create an unreasonable risk of harm to others or their property.[4] Subsequent cases reveal several activities that trigger strict liability, including blasting with dynamite and keeping wild animals. More recently, strict liability has also been applied to some pollution-generating activities.[5]

---

4. In fact, a classic case cited in the development of strict liability involved an environmental wrong. See Rylands v. Fletcher, Law Reports 3 House of Lords 330 (1868): the defendant was held strictly liable for nonnatural use of a reservoir on his land after water from the hilltop reservoir flooded onto the plaintiff's property.

5. For example, in Cities Service Company v. State of Florida, 312 So.2d 799 (Fla.App. 1975), a Florida District Court of Appeals held Cities Service liable without regard to negligence or fault for the damage caused by the accidental release of phosphate slime into a river.

Yet although the common law provides several avenues for forcing polluting firms to internalize their externalities, it has proved to be an ineffective constraint on excessive pollution.[6] There are several explanations for the shortcomings of private litigation. One explanation is the free-rider problem: many individuals attempt to receive the benefits of others' efforts without bearing any of the costs of private litigation. A second explanation is that the judicial system has neither the technical expertise nor the resources necessary to monitor and enforce its rulings in private nuisance cases. An additional explanation for the shortcomings of private litigation in controlling externalities involves problems in establishing the plaintiff's case. In many private nuisance actions, plaintiffs face considerable obstacles in establishing the necessary proof that the defendant's emissions are actually causing the alleged damages. For example, if more than one plant is emitting the same type of pollutant, it is almost impossible for a plaintiff to prove that one of the defendants caused the damages.

Nevertheless, many of the problems with basing claims for environmental harm on the common law could be solved with creative changes within the common-law tradition. For example, federal and state statutes could provide for pro rata liability for harm.[7] Class action and contingent attorney's fees are possible solutions to free-rider problems when large numbers of parties are affected. Recognizing new types of property rights and granting new causes of action to states could provide adjudicated solutions at the state level rather than regulations imposed from the federal level.

---

6. For example, in Boomer v. Atlantic Cement Co., 257 N.E. 2d. 870, (N.Y. 1970), the investment in the plant of more than $45,000,000 and its position as employer of more than 300 people outweighed the damage it caused. The dissent argues that this is essentially a license to pollute.

7. The technology for using isotopes to track pollution is rapidly developing to the point where it might soon be possible to identify more precisely the source of pollutants. This technology could be used to avoid some of the problems associated with market-share liability in products-liability cases where the identity of the manufacturer cannot be determined. See Hymowitz v. Eli Lilly and Co., 539 N.E. 2d. 1069 (N.Y. 1989), cert. denied, 107 L. Ed. 2d. 338; where there were many producers of the drug DES but no method to identify any firm as "the one" responsible for damages, a market-share theory of liability was adopted.

Early State Regulation and the Emergence
of Federal Regulation

For two interrelated reasons, legislation did not begin to displace common-law responses to environmental problems until long after the industrial revolution had brought about pollution on a large scale.[8] First, the scientific connection between pollution and public health was not established until many years after the development of indus-trial pollution. Second, and as a consequence, there were no well-defined interest groups lobbying for pollution-control measures. As scientific evidence began to make the case for environmental regulation, the states were the first to respond because localized interest groups formed to demand corrective legislation, often in response to litigation.

Donald Elliott, Bruce Ackerman, and John Millian have argued that a political imbalance led to the development of a peculiar form of state legislation aimed more at transferring wealth from out-of-state businesses to local environmentalists than at reducing pollution in a responsible way:

> The period of political cost-externalization . . . is charac-terized by the formation of organized groups of environmen-talists at the state and local level. Industry, however, remains passive and disorganized with regard to pollution issues. Politicians respond to the strategic imbalance cre-ated by the local organizational successes of environmen-talists by passing laws which place the primary costs of pollution control on out of state interests.[9]

It did not take long, however, for industry groups to begin to offset the environmentalists' victories at the state and local levels.

Once industry groups became active in defending their interests in pollution issues, they focused primarily on influencing federal regulations to support industrial advancement. Federal environmen-tal regulation evolved at a relatively slow pace until the late 1960s. By then there was a widespread belief that private litigation, state

---

8. See E. Donald Elliott, Bruce A. Ackerman, and John C. Millian, "Toward a Theory of Statutory Evolution: The Federalization of Environmental Law," *Journal of Law, Economics, and Organization* 1 (1985): 313, 316.

9. Ibid., p. 316.

antipollution programs, and some early federal legislation were inadequate protections for the natural environment. Moreover, there was a strong, rapid change in citizens' preference for a cleaner environment in the late 1960s.[10] Federal legislators recognized their constituents' developing concerns, and they responded more quickly than did their state counterparts. Once the stage was set, the current federally dominated regulatory structure emerged at a very rapid rate:

> This comprehensive structure of environmental regulation by the federal government is a curious feature of American law for at least two reasons. First, it developed fairly suddenly, seemingly out of nowhere. For two centuries, the effects of industrial pollution on the natural environment had been generally free from regulation by government, except for sporadic nuisance actions under the common law and a few municipal ordinances to control smoke. Second, it is curious that the environmental law of the 1970s was made primarily at the national level, rather than by the state or municipal governments which had traditionally held legislative authority over such matters.[11]

In essence, the growing public awareness of environmental issues presented entrepreneurial legislators in Congress with an ideal opportunity to take credit for strong environmental regulations.[12]

---

10. See Ladd and Bowman, *Attitudes toward the Environment*, pp. 3–7.

11. Elliott, Ackerman, and Millian, "Statutory Evolution," pp. 317–18.

12. Elliott, Ackerman, and Millian described this political dynamic as follows:

> Some of the time, at least, our polycentric lawmaking system has very different structural implications: rather than delay federal legislation during a lengthy period of experimentation on the state level, the federal system and the difficulty of organizing interest groups on a national level may sometimes encourage rapid and extreme lawmaking. Instead of checking and balancing opposing forces, the separation of powers may generate a system in which lawmakers compete to impress a poorly informed public with the strength of their symbolic commitments. Rather than prompting extended deliberation and broad consensus, our polycentric system may emphasize the strategic manipulation of passing organizational advantages and emotive symbolisms. The dynamic is this: environmental victories on the state level precipitate the counter-organization of certain specific polluters on the national level, which channels the legislative activities of credit-claiming politicians in the direction of preemptive federal lawmaking.

Ibid., pp. 328–29.

11

The Current Regulatory Framework

The first major effort by Congress to take the lead in combating pollution was the passage of the National Environmental Policy Act (NEPA) of 1969, which was signed into law by President Richard Nixon on January 1, 1970. NEPA is more a policy statement than a strict federal law intended to correct all environmental problems. It declares the continuing national environmental policy to be the following:

> [The federal government is] to use all practicable means and measures, including financial and technical assistance, in a manner calculated to foster and promote the general welfare, to create and maintain conditions under which man and nature can exist in productive harmony, and fulfill the social, economic, and other requirements of present and future generations of Americans.[13]

As implemented, NEPA policy has three major aspects: (1) it established the Council on Environmental Quality (CEQ); (2) it required all federal administrative agencies to evaluate environmental concerns before undertaking agency actions; and (3) it required federal agencies to prepare detailed environmental impact statements (EIS) before undertaking "major federal actions significantly affecting the quality of the human environment." NEPA and the CEQ are concerned with all aspects of environmental quality: wildlife preservation, land use, pollution, and even the effects of population growth on the environment.

The most important effect of NEPA has been on the behavior of federal agencies, since the purpose of the CEQ is to coordinate their environmental effect. In this regard, the EIS was a major innovation in determining the costs to the environment of certain governmental or government-approved activities. NEPA does not, however, give any special rights to individuals. For example, it does not give people the right to sue the government (or anyone else) when governmental activities pollute the environment. NEPA thus shows how the federal government was already moving away from the common-law approach, seeking instead to impose a new set of statutes and regula-

---

13. National Environmental Policy Act of 1969, 42 U.S.C. 4321.

tions on the problem. Although in a general sense NEPA is simply a policy statement, without the ability to regulate any behavior other than that of the federal government, the policy points toward increasingly greater federal control over environmental law issues.

In another major development toward the current regulatory structure, Congress created the Environmental Protection Agency (EPA) in 1970 as part of an executive reorganization plan submitted by President Nixon. The agency is headed by a single administrator, who serves at the pleasure of the president. The EPA was intended to streamline and strengthen the federal government's environmental protection policy through the consolidation of the pollution-control functions of several federal agencies. The agency administers the federal laws governing air pollution, water pollution, land pollution, noise, toxic substances, and pesticides. It establishes and enforces standards, conducts research on pollution effects, monitors and analyzes the environment, and assists state and local governments in their pollution-control programs. The EPA's powers of control over different types of pollution vary, depending on the authority granted it in particular legislation. The EPA complements the functions of the CEQ, advising it, for example, of new policies to protect the environment from pollution.

In its early years, the EPA adopted very high standards for pollution abatement.[14] The stringent regulations appeared to reflect the erroneous belief that the most efficient level of pollution is zero. As a result, the EPA was criticized for setting standards that were so high that the compliance costs to businesses far outweighed the benefits to society from reducing pollution.[15] Congress responded to these

---

14. The Clean Air Amendments of 1970 challenged the expertise of the EPA in finding an efficient method of pollution reduction. Congress required the EPA to set specific air-quality targets that "protect the public health" and required that such reductions be accomplished by 1977. See Bruce A. Ackerman and William T. Hassler, *Clean Coal/Dirty Air: Or, How the Clean Air Act Became a Multibillion-Dollar Bail-out for High-Sulfur Coal Producers, and What Should Be Done about It* (New Haven: Yale University Press, 1981), pp. 8–9.

15. This result is not at all surprising once one considers the incentives of regulators:

The single-subject agency, such as the EPA, proceeds with virtually no attention to the values that regulated activities might create apart from pollution. And, by statute and by practice—as well as by the inherent

13

criticisms by requiring the EPA to use some form of cost-benefit analysis in the setting of pollution emission standards. Cost-benefit analysis requires the balancing of costs and benefits of an activity to determine whether or not the activity—or the intensity of the activity—is justified.

During the 1970s, various federal governmental programs aimed at controlling specialized types of pollution were enacted or strengthened:

> This network of national statutes—together with a much larger body of implementing regulations promulgated by the Environmental Protection Agency—now constitutes one of the most pervasive systems of national regulation known to American law. Today, every discharge into the land, water or air—from the smallest smokestack to the largest landfill for the disposal of toxic chemicals—requires direct or indirect permission from the federal government.[16]

The extensive federal regulatory structure requires and allows for state involvement in pollution control. Although states still occupy a major role in implementing federal policies, this role is merely an administrative function designed to avoid some of the administrative diseconomies of federal regulation.

States are permitted to enact environmental statutes that are stronger than the federally imposed mandate.[17] In recent years, for example, many states have adopted environmental standards that are more stringent than the corresponding federal standards.[18] California is well known for taking the lead in requiring more stringent automo-

---

tendency of bureaucracy—timely responses to other values are impossible, even when such values are recognized.

Peter H. Aranson, "Pollution Control: The Case for Competition," in Robert W. Poole, Jr., ed., *Instead of Regulation: Alternatives to Federal Regulatory Agencies* (Lexington, Mass.: Lexington Books, 1982), pp. 339–93; quoted from p. 357.

16. Elliott, Ackerman, and Millian, "Statutory Evolution," p. 317.

17. See Wisconsin Public Intervenor v. Ralph Mortier, 501 U.S. 597 (1991).

18. See, for example, "State Air Board Tightens Rules on Burning Hazardous Waste as Fuel," *Environmental Law Reporter* (BNA) (May 17, 1991): 139 (Texas standards); and William R. Lowry, *The Dimensions of Federalism: State Governments and Pollution Control Policies* (Durham: Duke University Press, 1992).

bile-pollution abatement equipment. In 1991, nine northeastern states agreed to impose similar standards in their region.[19] Yet states are preempted from experimenting with statutes of less stringent standards. The current structure of environmental regulation provides a federal floor below which states cannot go, regardless of the preferences of their citizens. Perhaps more important, the states are not free to experiment with alternative implementation policies. Federal environmental policies appear to be stuck with an implementation philosophy that wastes billions of dollars each year.

---

19. See "Northeast, Mid-Atlantic States Reach Pact on Low-Emission Cars, Reformulated Gasoline," *Environmental Law Reporter* (BNA) 22 (Nov. 1, 1993): 1,643.

3

The Case for Federal Regulation

Federal domination of environmental regulation came about in part because environmentalists and others articulated persuasive arguments in favor of federal control. The alleged benefits of centralized federal regulation are related to the ability of the federal government to engage in activities and adopt policies that are beyond the scope of individual state activities and policies.[1] In this chapter, we summarize and assess the leading arguments in favor of federal regulation.

---

1. Professor Richard B. Stewart has stated the general case for national regulation:

> Markets alone cannot be relied upon to resolve many of the environmental, health, safety, and consumer problems created by industrialization and mass marketing. Moreover, state and local governments cannot deal effectively with these problems of market failure in the face of economically integrated national markets, products and capital mobility, and the rise of large multi-state businesses. . . . National measures are thus required to deal with problems generated by a national economy.

Richard B. Stewart, "Madison's Nightmare," *University of Chicago Law Review* 57 (1990): 335, 352. It is difficult to conceive of a statement more antithetical to the standard model of federalism. The federalism model is based on strong assumptions about political incentives, which in turn are driven by competitive forces. In contrast, Stewart's case for national regulation is based on a policy of granting a monopoly to national legislators and then trusting them to "do good." One should not forget that monopoly is also a market failure. Moreover, Professor Stewart's statement of the problem tends to redefine every local problem as a national problem for the simple reason that local lawmakers will have to consider how people in other jurisdictions might react. This in turn transforms every national problem into an international problem.

16

Limiting Interstate Externalities

The presence of interstate externalities means that states with pollution sources will not take all costs into account when formulating their environmental policies. The optimal state regulation, which would control pollution up to the point where marginal benefit equals marginal cost, would allow more pollution than would be optimal if all costs were internalized in the state's political process.[2] The neighboring state or states must bear the costs of pollution coming from outside. This situation is analogous to the primary justification of all environmental regulation—that is, forcing decision makers to bear or internalize the full costs of their decisions.

Acceptance of the interstate-externalities justification for federal environmental regulation does not dictate a specific type of regulatory response. The current regime of command-and-control regulations is no more justified under this analysis than are alternative, market-based approaches, such as the property-rights framework suggested in the next chapter of this volume. Rather than having federal regulators impose regulations on polluters, the interstate-externalities problem can be addressed by reallocating environmental authority in a manner that would force states and state decision makers to bear the full costs of their decisions regarding the regulation of pollution.

Halting the "Race to the Bottom"

A leading rationale for federal domination of environmental regulation is to prevent states from competing for economic growth opportunities by lowering their environmental standards in a so-called race

---

2. An alternative way of stating the problem is that state decision makers are unwilling to eliminate the externality because all the costs are borne in their state and some of the benefits accrue to citizens in other states. See, for example, Richard L. Revesz, "Rehabilitating Interstate Competition: Rethinking the 'Race-to-the-Bottom' Rationale for Federal Environmental Regulation," *New York University Law Review* 67 (1992): 1,210: "The other prominent market failure argument for federal environmental regulation is that, in the absence of such regulation, interstate externalities will lead states to underregulate because some of the benefits will accrue to other states."

to the bottom.[3] Presumably, all states would compete for economic growth by lowering environmental standards below the level they would select if they acted collectively at the national level. What is individually rational for individual states is collectively irrational at the national level.[4] Again Richard Stewart describes the implication of this dynamic in concise terms:

---

3. For thorough documentation of the influence of this argument, as well as a devastating critique of this argument, see ibid., pp. 1,233–44. Revesz stressed the importance of separating the interstate externality and race-to-the-bottom rationales for federal regulation:

> The distinction between the race-to-the-bottom and interstate externality rationales is critical for determining the appropriate scope of federal regulation. The concern over interstate externalities can be addressed by limiting the amount of pollution that can cross interstate borders, thereby "showing" upwind states the costs that they impose on downwind states. As long as the externality is eliminated, it would not matter that the upwind state chooses to have poor environmental quality—a central concern of the race-to-the-bottom advocates. Conversely, one could imagine a situation in which the environmental quality in the upwind state was very high, but in which there is nonetheless a serious externality problem because the sources in the states have tall stacks and are located near the interstate border, so that their effects are felt only in the downwind state.

Ibid., pp. 1,222–23.

4. This conclusion would hold even if there were no interstate externalities of the type described in the preceding section. The presence of interstate externalities and jurisdictional competition for economic growth are necessary for the competition to degenerate into a "tragedy of the commons," the common-pool problem. Such common-pool problems arise when a large number of firms, individuals, or other economic entities—such as states—all consume a single, finite, jointly owned resource at a faster rate than a single owner of the resource would use it, and the resource is unable to replenish itself. Thus, for example, if 100 people each own a single cow, and all 100 cows graze unrestrictedly in a single jointly owned field, the field's grass will be exhausted far more quickly than if the field had a single owner, because unlike a single owner, none of the 100 cowherds has any incentive to conserve or replenish the field's resources. In this regard, the environment can be viewed as a common pool that is "overgrazed" by states competing for economic growth. The tragedy of the commons requires two distinct conditions: interstate externalities and jurisdictional competition. Both interstate externalities and jurisdictional competition have been used as separate arguments in support of federal regulation. Hence we are treating them as separate arguments. Other commentators have combined the two arguments into a single tragedy-of-the-commons justification for federal intervention. For example, Richard Stewart has implied that the "characteristic insistence of federal environmental regulation upon geographically uniform

Given the mobility of industry and commerce, any individual state or community may rationally decline unilaterally to adopt high environment standards that entail substantial costs for industry and obstacles to economic development for fear that the resulting environmental gains will be more than offset by movement of capital to other areas with lower standards. *If each locality reasons in the same way,* all will adopt lower standards of environmental quality than they would prefer if there were some binding mechanism that enabled them simultaneously to enact higher standards, thus eliminating the threatened loss of industry or development.[5]

Thus, according to this logic, federal regulation is necessary to correct a political market failure at the state level. But there is a faulty link in the syllogism—*each locality does not reason identically*. Localities have different preferences for environmental quality, for a variety of economic and aesthetic reasons; it is by no means clear that competition between jurisdictions will lead to a lower level of environmental quality than would be held by a national median voter model.[6]

---

standards and controls strongly suggests that escape from Tragedy of the Commons has been an important reason for such legislation." Richard B. Stewart, "Pyramids of Sacrifice? Problems of Federalism in Mandating State Implementation of National Environmental Policy," *Yale Law Journal* 86 (1977): 1,196, 1,212. Most of Stewart's tragedy-of-the-commons argument is really a race-to-the-bottom argument that does not depend on the existence of interstate externalities. In fact, Revesz cites Stewart's argument as a race-to-the-bottom rationale. Revesz, "Rehabilitating Interstate Competition," p. 1,210.

5. Stewart, "Pyramids of Sacrifice?" p. 1,212 (emphasis added).

6. Revesz makes the same point:

Finally, it is important to stress that the existence of interstate competition for industry is not sufficient, by itself, to produce a race to the bottom or, consequently, to justify federal regulation. Obviously, a race to the bottom requires not just a "race," but also that a competitive jurisdiction adopt a less stringent pollution control standard than an otherwise identical island jurisdiction would have adopted. Second, it requires that the less stringent standards that emerge from the competitive process be socially undesirable. Otherwise, the case for federal regulation disappears, or, alternatively, federal regulation must be justified on a different basis.

Revesz, "Rehabilitating Interstate Competition," p. 1,219.

In fact, competition between jurisdictions may lead to greater increases in environmental quality. It is often argued that environmental quality is a luxury good, in that individuals develop a greater concern for environmental issues as their incomes rise.[7] If this is true, the key to increases in environmental quality may be found in higher incomes. This point has implications for the desirability of jurisdictional competition, as illustrated by the following example.

Assume that there is no national environmental regulation, and all environmental issues are the prerogative of the state and local governments. Firm X operates in New Jersey. As the incomes of those who live in New Jersey increase as a result of industrial growth provided by X, the citizens of New Jersey will place a higher emphasis on environmental quality. State and local government decision makers will respond to citizens' demands for better pollution control. Assume that X responds to the increase of pollution standards in New Jersey by moving to Missouri, where pollution control is not as stringent. Missouri's environmental laws could reflect Missourians' preferences, given their relatively low incomes. Many people in Missouri welcome X's operations, even at the expense of environmental problems. As X's industrial production causes Missouri's economy to expand, the incomes of individuals will increase, and so will their demand for a cleaner environment. The initially harmful levels of pollution create the minimum levels of wealth necessary to fuel citizens' demands for a cleaner environment. The competition among different states may enhance economic growth and accelerate the evolution of more efficient pollution abatement equipment.[8]

---

7. Studies indicate that environmental awareness begins at an income level of $5,000. See Bruce Yandle, "Is Free Trade the Enemy of Environmental Quality?" in *NAFTA and the Environment*, Terry L. Anderson, ed. (San Francisco: Pacific Research Institute for Public Policy, 1993), p. 9; citing Gene M. Grossman and Alan B. Krueger, "Environmental Impacts of a North American Free Trade Agreement," working paper no. 3,914, Cambridge, Mass.: National Bureau of Economic Research, November 1991. Studies also found that Communist countries have systematically higher levels of pollution, all else equal.

8. As incomes rise, a cleaner environment will become a good that more people demand. Over time firms would search for new pollution abatement technology instead of moving from state to state. As different jurisdictions identify their proper allocation between environmental quality and economic growth, X will have the greatest incentive to develop efficient pollution-control technologies. At this point

Finally, the race-to-the-bottom rationale for federal government domination of environmental regulation is based on the assumption that the federal government can in practice do a better job than can the state governments. There are strong reasons to believe that this assumption is wrong. The race-to-the-bottom justification for federal intervention, while critical of state political processes, ignores the problem of interest-group domination of the legislative process in Washington. The interest-group problem is more acute at the federal level than at the state level because of the lack of competition among regulators at the national level.[9] There are numerous reasons, however, to believe that the Washington political market reflects its own regulatory common-pool problem, with politicians logrolling between environmental votes and votes on unrelated issues. Unfortunately, the race-to-the-bottom rationale underlies much of the federal environmental statutes.[10]

Controlling Political Cost Externalization

State environmental regulations that impose financial costs on out-of-state producers are often cited as a justification for federal intervention.[11] Some state environmental regulations restrict local consump-

---

any increase in production cannot have a corresponding increase over the optimal level of pollution. In order to increase production at this point, more efficient technologies must be developed. A dynamic view of the economy sees competition creatively replacing lower-valued, inefficient producers with higher-valued, efficient producers. Current levels of pollution are simply a temporary situation that will be removed as our economy grows and individuals' incomes increase. See ibid., pp. 1–10 using the analysis above in the context of U.S.-Mexican relations under NAFTA.

9. Of course, Richard Stewart is well aware of the influence of interest-group politics on environmental policies. In fact, one of the items he lists as a possible rationale for centralization of environmental regulation is that environmental groups are likely to have relatively greater influence in Washington than in the states. Stewart, "Pyramids of Sacrifice?" p. 1,213. Although this may have been true at some point, it may not be true at all times. See Elliott et al, "Statutory Evolution." Moreover, it is not at all clear that greater influence for self-styled environmentalists is the best policy for the environment.

10. See Revesz, "Rehabilitating Interstate Competition," p. 1,212.

11. Elliott et al., "Statutory Evolution," p. 329 suggest that:

Federalism opens up the possibility of a distinctive credit-claiming strattion of products produced in other jurisdictions. The classic modern example of cost externalization is California's strict automobile-emissions–control policy. But the concern here is that in enacting legislation, the local legislators ignore the regulatory costs imposed on out-of-state automobile manufacturers who are unable to pass all the cost increases on to consumers in the regulating state. In effect, it is alleged that political cost externalization is a political market failure that requires federal regulatory intervention.

Even if the cost externalization analysis is correct, the implications of the analysis for the structure of federal regulation are not obvious. Historical experience suggests that caution is called for in responding to cost-externalization problems.[12] Thus the federal response should address the cost-externalization problems in the least restrictive manner. Federal regulations that preempt stringent local environmental regulation of local externalities may be justified on the ground that the local regulations impose tremendous costs on businesses' national marketing strategies. Several possible solutions to this economic problem, however, avoid federal preemption and thus allow for the achievement of some of the benefits of federalism. First, the federal government could impose maximum limits on state regulations that affect products manufactured in one state but sold in another. States would be free to set environmental standards up to, but not above, this level. The problem with this approach is that the larger states adopt the maximum standard, and the maximum becomes a minimum requirement.

A second possible solution to the alleged political cost-externalization problem is that the federal government could prohibit individual states from mandating design changes in products manufactured

---

egy for aspiring politicians on the state level, which we call *cost-externalization*. Quite simply, dividing the nation into fifty geographic zones makes it almost inevitable that some pollution problems will be generated by out-of-staters. Since Midwestern autoworkers don't vote on whether California should ban the internal combustion engine to control smog and Appalachian coalminers don't vote on whether New York should ban coal to control sulfur oxides from power plant smoke stacks, these issues promise politicians on the state level the equivalent of a free lunch— "tough" legislation allows them to garner public credit for bringing a benefit to *their* constituents at somebody else's expense.

12. Recall that Elliott et al. argued that the cost externalization problem was one of the initial catalysts for federal regulation. See ibid.

in other states. State responses could be limited to the least restrictive policy in terms of adverse consequence on national marketing strategies. Take, for example, the Maine statute that prohibits the use of a particular type of fruit-juice container because it is neither biodegradable nor recyclable.[13] An alternative policy that would result in less disruption of the fruit-juice manufacturers' distribution systems would be a corrective tax on the containers. Such taxes would have to be structured so that the level charged corresponded to the level of local pollution caused by the product. Because a large portion of the tax would be borne by the local consumers, local politicians would face a greater constraint in setting the taxes than they do in setting pollution standards when they can externalize the political costs.[14] Of course, the obvious problem with allowing federal regulations to restrain state activities is that it could result in a cure worse than the disease.

Furthermore, the presence of political cost externalization does not mean that there has been a political market failure. California's decision to require the installation of expensive antipollution equipment on all new cars sold in the state adds to the marginal cost of producing the cars sold in California. As such, the increased marginal cost is analogous to a per car excise tax in its effect on the selling price of automobiles in California. The incidence of the regulatory requirement is the same as the tax incidence of a per car excise tax. The marginal cost of the pollution equipment is shared by California consumers, who must pay more for cars, and by out-of-state manufacturers, who receive a lower after-regulation price because of the increased marginal costs. To the extent that California

---

13. A possible explanation for the Maine legislation is that it places fruit-juice sellers at a competitive disadvantage relative to their competitors.

14. Of course, some of the costs under political cost externalization are borne locally in the form of higher prices, but they are less obvious to the consumer than would be pollution-based excise taxes. A potential problem with this approach is that legislators may be tempted to use Pigovian taxes as revenue sources, unrelated to the correction of real environmental problems. In order to prevent this type of barrier to interstate commerce, states should be required to use Pigovian tax revenues solely for environmental projects that serve to diminish demonstrable externalities. Legislators' decisions would focus on the correction of externalities and not on revenue generation. That is, the benefit from the tax is reduced pollution, not simply increased general revenue to the state.

consumers observe that they must pay more for new cars than do consumers in bordering states, the costs to California consumers are taken into account by California legislators.

Moreover, the costs imposed on out-of-state manufacturers cannot be ignored by state legislators, because the out-of-state manufacturers will make political contributions, hire lobbyists and public relations firms, and otherwise attempt to prevent the passage of the legislation. It is naïve to expect out-of-state firms to accept the huge costs of the regulations passively. The out-of-state interests may even have an advantage over environmental groups in organizing their involvement in California's political process.[15] Finally, because the higher prices caused by the regulations will result in fewer new car sales, new car dealers will have incentives to lobby California legislators not to adopt the regulations.

The fact that a particular cost-externalizing regulation is adopted does not mean that the adopting legislators ignored the out-of-state costs; it means simply that the legislators decided the benefits were greater than the costs. This analysis suggests that the political cost-externalization justification for federal environmental regulation is not a valid justification for federal intervention. There is no political market failure. But even if there are some problems in the political market, they are likely to be small as compared with the problems of centralized federal regulation.

Capturing National Economies of Scale in Administration, Technical Expertise, and Funding

It is often asserted that state and local regulation is inadequate because states and localities usually lack sufficient administrative and enforcement resources. Once again, our analysis returns to the effects of a political decision: if the external costs that are allowed by the inadequate administration and enforcement resources are purely local, then the failure of the local politicians to allocate resources to deal effectively with the pollution is a local problem. Moreover, the federal government does not have the resources to resolve the problem on its own:

---

15. The defeat of California's "big green" proposition was attributed to the role of out-of-state interests in supporting opposition to the stringent regulations.

24

The political obstacles to congressional creation and funding of a massive federal inspectorate and police force adequate to the task appear insurmountable. Even if such a force were created, federal environmental goals could not be achieved without the cooperation of state and local authorities with responsibility for water supply, highway location, traffic control, mass transit, land use planning, and other governmental programs related to environmental management.

The inadequacy of federal resources in comparison to the magnitude of environmental problems inevitably results in federal dependence on state and local authorities. Often federal air and water pollution control statutes give the states initial responsibility (subject to federal review and "back-up" enforcement) for achieving federal objectives. In other instances, the EPA is authorized to delegate certain of its own implementation and enforcement responsibilities, an option which overburdened federal officials have readily utilized. Even where no formal delegation has occurred, the EPA in practice relies heavily upon the cooperation of state officials.[16]

Thus, in light of Congress's unwillingness to provide funds to solve environmental problems, it is illogical to assert that federal intervention is necessary because the states' funding of their environmental agencies is deficient.

Everyone wants a cleaner environment if it is free, but neither the politicians in Washington nor those in the state capitals are willing to come up with the necessary funding. Politicians in Washington should not be given credit for their deep concern about the environment unless they actually provide the funding for the programs they mandate.[17] Washington has not provided the necessary

---

16. Stewart, "Pyramids of Sacrifice?" pp. 1,200–1.

17. This is especially true in light of the observation that centralization obscures the true costs of environmental and other policies:

Centralization also makes less apparent the sacrifices involved in public expenditure to promote environmental quality. The relation between one's tax payment into the large and complex federal fisc and any particular federal expenditure is obscure; the correlation between a state or local bond issue for sewage treatment facilities and personal financial sacrifice is more direct and immediate. The ambitious municipal waste treatment

funding. It appears rather that the economies-of-scale justification for centralization of environmental regulation is political grandstanding, not actual funding and administration of programs.

One real source of economies of scale associated with centralization of environmental regulation could be in centralized research on technical, scientific issues that recur throughout a number of different states. Much of the information generated in this process is a public good that is best provided by government funding. Similarly, centralization of data collection and dissemination is likely to be a cost-effective technique of identifying trends across states and setting policy priorities. These economies can be realized by the federal government even when most policy-making and implementation functions are handled by the states.

---

programs adopted in federal legislation would probably have been rejected in many states and localities.

As noted above, the federal health and environmental protection bureaucracies are generally larger and more professional than their state and local counterparts. Once a substantial program of environmental protection is launched, these federal bureaucracies' very size, professional orientation, and remoteness also make them comparatively less sensitive to public discontent when the economic and social costs of such programs become apparent, particularly if these costs fall disproportionately on a few regions. For analogous reasons, public protests, especially if localized, will have less impact on federal judges and legislators than on their state and local counterparts.

Thus, a variety of "ratchet" factors make it less likely that federal (as opposed to state or local) environmental programs initially undertaken in part out of moral concern will be abandoned or compromised because of the sacrifices entailed. Under centralized decisionmaking these sacrifices may be less visible (because of fiscal mechanisms) or more palatable (because widely shared). Or the sacrifices may be discounted because federal officials are simply less sensitive to short-term swings in public attitudes. These features of national decisionmaking would be welcomed by those who embrace a genuine moral commitment to environmental protection but fear their inability to maintain that commitment in the face of subsequent privations. Delegation of environmental programs to the federal government can accordingly be viewed as a self-binding mechanism—an insurance policy against *akrasia*. The emphasis in federal programs on wilderness and species preservation, on uniform health-based pollution control standards, and even the extravagant zero-discharge goal of the 1972 FWPCA all reflect the non-utilitarian moral and sacrificial aspects of environmental policy.

Ibid., pp. 1,218–19. Some may view this as an indictment rather than a virtue of centralization.

Finally, whatever the economies of scale associated with centralization of environmental policy, they are surely overwhelmed by the diseconomies of scale in centralized administration. There are several hundred thousand industrial sources of air and water pollution and more than a million hazardous-waste generators in the United States.[18] The enormous job of regulating these pollution sources is compounded because there are many different types of sources and tremendous differences in local environmental variables. The environmental harm caused by the emission of the same amount of pollution can vary widely, depending on local environmental conditions. For example, the discharge of polluted water into a large body of water may have no discernible effect, while the same discharge into a small, pristine stream may have disastrous consequences. Federal regulators never have been and never will be able to acquire and assimilate the enormous amount of information necessary to make optimal regulatory judgments that reflect the technical requirements of particular locations and pollution sources.[19] Rather, regulators have responded to the information problem by imposing uniform, technical standards.[20]

The diseconomies of scale in environmental regulation make pollution control much more expensive than necessary:

> If controls were tailored to individual plant costs, our current expenditures for air and water pollution control could be reduced from over $50 billion annually to $25 billion or less with no sacrifice of overall environmental quality. Such tailoring, however, is an administrative impossibility in a centralized system of regulation. Federal administrators in Washington could not possibly devise individually-tailored standards for each of hundreds of thousands of plants and

---

18. Stewart, "Madison's Nightmare."

19. The analogy to information processing in a market economy versus centralized planning is obvious. Markets generate prices that convey incredible amounts of information about individual preferences at any time or place. See F.A. Hayek, "The Use of Knowledge in Society," *American Economic Review* 35 (1945): 1.

20. These problems with centralized command-and-control environmental regulation are also present in other areas of federal regulation. See Stewart, "Madison's Nightmare," p. 343.

facilities, particularly when each standard would require a formal hearing and a possible lawsuit.[21]

The national diseconomies of scale swamp the alleged and illusive national economies of scale as soon as national regulation is instituted.

State regulations, of course, would be subject to similar problems. The question then becomes whether the state governments can do a better job than can the federal government. State regulators are faced with similar information problems in setting individually tailored standards, albeit on a smaller scale. The high administrative costs of setting such standards may lead state regulators to set uniform, statewide standards, where the higher costs of pollution control are not reflected in their administrative costs. Nevertheless, there would appear to be considerable gains available from tailoring regulations at the state rather than the federal level, even if the result is to have uniform technical regulations within each jurisdiction.

The primary justification for federal environmental regulation is the control of extrajurisdictional externalities, but this does not require that the federal regulators attempt to micromanage state environmental responses. Indeed, federal regulators appear to be incapable of doing the job. The economies of scale in federal regulation are illusory. The federal government's realization of economies of scale in administration and technical expertise does not come close to justifying federal preemption of local regulation of local externalities. And it is not clear that federal funding is more adequate for the tremendous task it faces than is state funding for the much smaller task it faces.

Maintaining National Moral Ideals

Another argument for federal regulation of even strictly local externalities is that the federal level is the one best suited to reflect the moral obligation of U.S. citizens to one another, as well as to future generations:

> In situations where pervasive and significant spillovers do not exist, it is necessary to consider the reserved question

21. Stewart, "Controlling Environmental Risks," p. 156.

of Congress' power to compel state cooperation in the name of national moral ideals, since the analysis has demonstrated other rationales for infringing state autonomy to be weak.

The case for federal intervention to help realize moral ideals, such as protection of susceptible minorities or the opportunities of future generations, is only somewhat less strong than the spillover rationale. These ideals are valuable not merely for their own sake but also for the moral education fostered by their consideration. Environmental problems force us to face consequences of our immediate actions that we would prefer to disregard because of their disturbing impact on fellow citizens, on future generations, and on the nature of our society. Such a confrontation is indispensable to the collective moral growth of our society. Given the logic of the "politics of sacrifice," this form of collective education is likely to be attenuated if the crucial decisions are excessively noncentralized.

It is not the case, however, that federal intrusions on local self-determination are justified so long as there is some moral purpose arguably served thereby. Three conditions should be met in order to justify use of the commerce power to coerce state implementation of national moral goals. First, the goals should be among those that could persuasively be regarded as basic in a reflective ideal of the good society. Second, the goals should be a sort that are unlikely, because of structural defects, to be realized under a regime of non-centralized decisionmaking. Third, federal intervention should promise a substantial contribution to the realization of the goals. As in the case of interstate spillovers, courts should undertake a substantial inquiry in determining whether these threshold conditions have been met. There would seem to be only three types of environmental goals that could meet these criteria: the prevention of serious harm to human health; maintenance of diverse environments to stimulate individual and collective cultural development; and preservation of irreplaceable environmental assets for future generations. In these aspects of environmental protection, as in the elimination of racial and sexual discrimination and the effort to provide all with minimum levels of well-being, effective capacity for central direction in the definition and realization of moral ideals will often be indispensable.[22]

We strongly disagree with this argument favoring centralized control of environmental policy making based on Mr. Stewart's particular morality rationale.[23] The moral-ideals justification for centralization is based on the flawed presumption that it is moral for the federal government to force people to pay for goods they do not want.[24]

There are also several practical problems with this argument for

---

22. Stewart, "Pyramids of Sacrifice?" pp. 1,264–65.

23. Ibid., pp. 1,217–18.

24. See ibid., pp. 1,221–22:

Moral crusades enjoy little credit with the nonbelievers who are taxed to underwrite such ventures. Motorists facing drastic curtailment of mobility, the poor with increased utility bills, and the unemployed in rural areas closed to development may understandably view the sacrifices they are called upon to make as excessive. Resistance and resentment may be heightened by the fact that many environmental programs distribute the costs of controls in a regressive pattern while providing disproportionate benefits for the educated and wealthy, who can better afford to indulge an acquired taste for environmental quality than the poor, who have more pressing needs and fewer resources with which to satisfy them. These circumstances may foster, and in part justify, a cynical attitude towards the moral justifications advanced by upper-middle class advocates for environmental programs which benefit that class disproportionately. The impairment of local political mechanisms of self-determination and official accountability involved in federally dictated environmental programs affords further grounds for resentment.

It is not too fine a conceit to mark a parallel between the local impact of national environmental policies and Peter Berger's assessment of the social and moral costs of development in third-world nations. In his book *Pyramids of Sacrifice*, Berger decries the insensitive willingness of governmental elites to impose severe sacrifices on the populace, repressing opposition to such sacrifices on the grounds that they are necessary to "development" but will not be undertaken voluntarily and that once development has occurred the society will look back upon the sacrifices as justified. Aspects of national environmental policy might similarly be viewed as the insensitive imposition of sacrifices on local communities, viewed as unjustified by those that bear them (in particular the poor communities), for the sake of a national elite's vision of a better society. Why should Washington force San Francisco to have cleaner air than it apparently wants?

This view also is reflected in Jack Kemp, "Free Housing from Environmental Snobs," *Wall Street Journal*, July 8, 1991, p. A6.

centralization. First, there is no reason to think that a centralized authority can deliver regulations that meet whatever moral ideals many of us share. For example, powerful arguments and evidence support that private-property owners do a much better job of preserving and protecting large tracts of land.[25] Government-controlled land is more likely to be spoiled than is privately held property, since the bureaucrats who control public land do not bear the costs of overuse; in fact, they obtain political support from interest groups in exchange for allowing such overuse.[26] Thus, even assuming that there is a strong public ideal that favors a cleaner environment, there is no reason to believe that centralized decision making is the best strategy for attaining that ideal.

An additional argument against the moral-ideals justification is that it is highly open-ended and indeterminate. Anybody can argue that his version of a particular law is more legitimate than his rival's, on the grounds that his is more consistent with the moral ideals of the nation. This argument is impossible either to refute or to prove. The most reliable guide for the moral ideals of a polity as diverse as the United States lies in the revealed preferences of its citizens—that is, in the willingness of its citizens to pay for environmental quality. Appeals to the moral ideals of the nation are often thinly disguised appeals to authority when more substantive policy justifications are lacking.

Whatever the benefits of centralization may be, centralizing authority over environmental policy has its costs. Local preferences for varying levels of environmental quality are ignored, and the laboratory of the states is destroyed. Moreover, centralization makes it very difficult to identify and correct the inevitable mistakes made by environmental policy makers. No one is prepared to argue that Congress is perfect, or that the Environmental Protection Agency is above the influence of interest groups and partisan politics. Environmental policy may be too lenient or too strict, and implementation may be wasteful, but there is no corrective mechanism once policy making is centralized. In fact, the "iron triangle" of congressional committees, government bureaucracies, and industry and environmental lobbying

---

25. See generally Anderson and Leal, *Free Market Environmentalism*.
26. Ibid.

groups is seen by many as conspiring to maintain the centralized status quo in the face of tremendous evidence that it is increasingly wasteful,[27] and of political theory describing why centralization was excessively ambitious in the first place.[28]

---

27. See Bruce A. Ackerman and Richard B. Stewart, "Reforming Environmental Law: The Democratic Case for Market Incentives," *Columbia Journal of Environmental Law* 13 (1988): 171, 172.

28. See Elliott et al., "Statutory Evolution."

4

The Efficient Allocation of Environmental Regulatory Authority

The extreme centralization of environmental regulation is the result of interest-group politics and dramatic political developments rather than the sober analysis of the major trade-offs involved in moving to federal domination of environmental protection. During the late 1960s and early 1970s, many environmental concerns previously seen as local issues, such as solid waste disposal, became federal issues. Certainly national politicians were ahead of the curve in responding to environmental concerns, but that does not mean that state and local politicians would have continued to be unresponsive to environmental issues in their own back yards. Nevertheless, there appears to be a type of one-way ratchet in the federal system—once a federal issue, always a federal issue. Moreover, among many environmentalists, there remains a deep distrust of state and local control of environmental quality. An analysis of the incentives of state and local politicians suggests that this distrust may not be justified.

There are numerous trade-offs in the allocation of environmental regulatory authority within a federal system. For example, widely different levels of interest in environmental quality across states lead to the development of a hodgepodge of state regulations, which creates confusion and inefficiencies in businesses' production and nationwide marketing strategies.[1] Yet there are problems with federal

---

1. The strict California air pollution regulations are but one of the many examples of this confusion. Thus it is not surprising that in recent years there has been a great deal of discussion among business groups about whether they are better off with state or federal environmental regulations. In general, business groups are in favor of federal preemption of the hodgepodge of state regulations.

preemption of state environmental regulations. First, federal preemption may reduce the ability and incentives of state regulators to experiment with creative solutions to local environmental problems. Second, federal preemption centralizes many environmental decisions in Washington, where interest groups dominate decision making. Thus economic consequences, particularly at the local level, are often ignored. Related to this ignoring of economic consequences is a third problem with federal preemption: it fails to provide sufficient funding for required local actions.

This chapter addresses ways to allocate regulatory authority so that political institutions and processes will yield policies that achieve the optimal or most efficient level of pollution without imposing unnecessary costs on productive economic activity.[2] There is no definitive answer to the critical question, "what is the optimal level of pollution?" Policy analysts must therefore assert that the determination of the tolerated amount of pollution is a political question. Although "invocations of the superiority of political processes for resolving issues of social and economic policy" are commonplace, the assertion that a policy issue is a political problem does not solve the policy issue, because alternative political institutions are likely to yield different answers.[3] Thus, alternative political policy-making institutions, such as state and national legislatures, can be analyzed to determine which institutions are best suited to weighing costs and benefits in the determination of the optimal level of pollution. To the extent that all relevant costs and benefits are taken into account in the political decision-making process, "better" policies should

---

2. Unfortunately, standard economic definitions of efficiency are often difficult to apply to environmental issues because of widely divergent views of the costs of pollution (or the benefits of pollution reduction). One definition of efficiency is straightforward: don't waste resources. Bruce Ackerman and Richard Stewart, for example, have shown that the adoption of a combination of pollution-based statutes with market-based incentives could achieve at least the *same* level of environmental quality at dramatically lower costs. Bruce A. Ackerman and Richard B. Stewart, "Reforming Environmental Law: The Democratic Case for Market Incentives," *Columbia Journal of Environmental Law* 13 (1988): 171. But determining the *optimal* level of environmental quality—where the marginal benefit of additional pollution abatement is equal to the marginal cost—is an entirely different matter from determining the lowest-cost way of achieving some identified quality level.

3. Stewart, "Madison's Nightmare," p. 348.

emerge. This leads to a comparison of the relative costs and benefits of federal and state regulation of environmental risks.

This chapter begins with an introduction to the economics of federalism, which focuses on the conditions under which competition between jurisdictions will produce optimal environmental laws and regulations. We then turn to a consideration of a variety of situations where the federalism conditions are satisfied to differing degrees. This analysis yields some fairly straightforward principles about which level of government should be granted regulatory authority to deal with different types of externalities.

Environmental Quality and Jurisdictional Competition

In most areas of economic activity, competition produces the efficient or optimal allocation of resources. It is at least plausible that competition among states for environmental quality may generate the optimal combination of environmental regulations across the country.[4] Competition among political jurisdictions is likely to generate optimal laws if four conditions are fulfilled:

1. The economic entities affected by the law must be able to move to alternative jurisdictions at a relatively low cost.
2. All the consequences of one jurisdiction's laws must be felt within that jurisdiction.
3. Lawmakers must be forced to respond to adverse events such as falling population, falling real estate prices, falling market share or revenue, and other manifestations of voter discontent that result from inefficient regulations.
4. Jurisdictions must be able to select any set of laws they desire.[5]

Of course, in the real world there are no purely local externalities, no perfect markets, and no perfect governments. As a consequence,

---

4. For a recent statement of this position, with a concise summary of the relevant literature, see Revesz, "Rehabilitating Interstate Competition," pp. 1,233–44.

5. This list represents our summary of the literature. The seminal article in this literature is Charles M. Tiebout, "A Pure Theory of Local Expenditures," *Journal of Political Economics* 64 (1956): 416. Important contributions include Frank H. Easterbrook, "Antitrust and the Economics of Federalism," *Journal of Law and Economics* 26 (1983): 23; Roberta Romano, *The Genius of American Corporate Law* (Washington, D.C.: AEI Press, 1993).

failure to achieve all four conditions is not a mandate for federal government intervention; it is only an indication that local regulation may be imperfect.

The first federalism condition requires that the economic entities affected by the law must be able to move to alternative jurisdictions at a relatively low cost. With respect to environmental protection laws and regulations, this condition applies to two types of economic entities. The first is the parties adversely affected by the pollution, such as individuals and households, as well as businesses that must pay higher wages to attract workers to a polluted area. The second type of economic entity comprises the polluters adversely affected by the high cost of compliance with the jurisdiction's environmental laws and regulations. Most economic entities are mobile, at least in the long run; and both types of entities will always contain a substantial portion of marginal entities that are very mobile. Regulators at both the state and the national levels, however, will be aware that some firms are more mobile than others. Firms lacking mobility are particularly vulnerable targets to governmental regulation that threatens to expropriate investments in immobile capital. In determining how regulatory authority should be allocated between state and federal authorities, we should be concerned about the ability of governmental actors to expropriate fixed investments.

Government appropriation of fixed investments can take a variety of forms. For example, an industry might spend considerable resources simply learning the details of a particular state's environmental law before entering that state to do business. The firm's investment in learning that law is a fixed investment that would be appropriated if the law changed or was preempted by Congress. Hence, beneficiaries of a particular regulatory regime might prefer to keep an existing regulatory structure in place even where a marginally superior alternative exists if the benefits of the new regulations are outweighed by the costs of learning to cope with the new regime.

Similarly, a firm might make a considerable investment in configuring plant and equipment in reliance on an assumption that a particular set of environmental laws will remain in place for a certain period of time. The ability of politicians to change—or threaten to change—the applicable environmental laws reduces the incentives of firms to make investments in capital assets. Thus, a sensible environmental policy will seek to allocate authority among state and federal regulators to reduce the possibility of expropriation of fixed capital investments by industry. This can be done only by establishing clear spheres of authority between state and federal actors and by limiting the incidence of overlapping regulatory authority. But, in general, if a state is able to expropriate capital because of the inability of a firm to leave the state, then the federal government is in an even more advantageous position, because it is usually more difficult to leave the country than to leave the state.

Although the discussion of the mobility condition focuses on exit from unfavorable jurisdictions, the entry of mobile economic entities to more favorable jurisdictions is also a significant factor. The two types of economic entities identified here may have conflicting preferences. For example, individuals affected by pollution are likely to favor strict environmental laws, while polluters are likely to favor lax environmental laws. The federalism model is designed to find the optimal balance between these conflicting preferences at the local level.

The second federalism condition requires that all the consequences of one jurisdiction's laws must be felt within that jurisdiction. Some types of local externalities involve the location of a stationary pollution source—for example, a factory—within the local jurisdiction. Where this is the case, the local political decision makers will take the costs imposed on the factory into account when devising local environmental policy.

This condition is violated where a state has lax environmental regulation, and pollution spills over from one jurisdiction to another. As discussed above, this interstate externality is a strong justification for some form of federal intervention. If the pollution allowed by one state's lax regulation crosses the relevant political boundaries, then there will be a political market failure, regardless of whether it is passed in the name of economic development.[6] Local governments can be prevented from playing this game by state regulations or policies, and states can be prevented by federal regulations or policies. The extent of the response could be fairly minimal when compared with today's command-and-control regulation. For example, states or even individuals could be given the right to sue neighboring

---

6. See discussion of definition of property rights among states, below.

states that fail to meet minimum federal standards.[7] Alternatively, the federal government could arbitrate claims between states involving interstate pollution.

The third federalism condition requires that the political process be sufficiently competitive so that lawmakers are forced to respond to adverse events. The basic federalism model is based on the belief that lawmakers enact laws reflecting local preferences as reflected in the political process. Thus, constituents who are not on the margin with respect to mobility or exit can still exert considerable influence by exercising their voice option in the political process.

State and local environmental regulation is often claimed to be inadequate because states and localities are under pressure to relax environmental controls in order to attract industry.[8] To the extent that such pressures influence environmental policy, many of the costs (negative externalities) and benefits (economic growth) are borne locally. If the pollution is purely local in all respects, then there is little justification for questioning the motivations of local politicians. Moreover, competition between states can be viewed as beneficial because it forces politicians to consider the costs as well as the benefits of environmental regulations:

> Some may object that state and local governments will compete for industries by offering lax environmental standards. We suspect that this is a very real possibility and welcome its effects. In particular, state and local governments will balance voters' interests in economic activity and environmental quality more closely than the federal government will. Therefore, a few states may offer themselves as sinks and sewers, but that will save the rest of the nation from these depredations. Similarly, many states may have residents who want a much higher environmental quality than federal regulations now mandate. This higher quality is more likely to prevail under local control.[9]

---

7. See, for example, text accompanying notes, below.

8. For a summary and critique of the race-to-the-bottom analysis, see Revesz, "Rehabilitating Interstate Competition."

9. Aranson, "Pollution Control," pp. 383–84. A concern about "sinks and sewers" under marketable permit plans suggests that some concentrations of pollution may not be especially harmful for most types of pollution:

Allowing for local decision making at least leaves the choice regarding a given level of pollution with the people most likely to be affected by it.

The fourth federalism condition is that jurisdictions must be able to select any set of laws they desire. If the first three federalism conditions are met, then the economics of federalism suggests that local governments should retain discretion in selecting the level of environmental quality they prefer as well as the regulatory policies used to achieve those goals. Granting this authority to local jurisdictions should generate benefits from several sources:

> First, public policy toward environmental policy would more accurately reflect the preferences of those affected. Second, where serious divergences from individual preferences do occur, people have the option of moving to more favorable locations. Third, a real interpolity competition in public policy toward the environment would emerge, as would productive experimentation with governmental alternatives. A veritable marketplace of governments would give the citizen, the consumer of public services, a choice among the competing units. Fourth, decentralization would also generate competition in the use of externality-abatement techniques. Fifth, decentralization would help to ensure that resources flow to their highest-valued use, because those who would receive the benefits of an improved environment would also have to pay the cost. Finally, by reducing substantially the number of people involved with particular environmental problems, decentralization would markedly diminish transactions costs, which, in turn, would allow for the use of market-like abatement policies. We should not deceive ourselves that state and local governors are better than EPA officials.

Charges and marketable permit systems are designed to induce an aggregate reduction in pollution or risk without ensuring a particular level of control at any given facility or location. It may therefore not be appropriate in dealing with pollutants or chemical risks that have localized "threshold" effects, causing serious damage only if they exceed a given concentration at a particular location. But many, perhaps most, environmental risks do not involve such thresholds.

Stewart, "Controlling Environmental Risks," p. 161.

> However, decentralization allows other people to visit on
> legislators and regulators the content of their preferences
> and the rigors of the marketplace.[10]

Thus, decentralization would encourage the adoption of the optimal pollution-abatement policies.

Although jurisdictional competition in environmental regulation is not perfect, it must be compared with the relevant alternative: federal preemption of state regulation with centralized, monopoly regulation. In fact, one of the primary benefits of federalism—the ability of states to experiment with new policies—is all but eliminated by centralization. Moreover, the political accountability that drives jurisdictional competition is replaced by necessary delegation of major legislative decisions to federal bureaucracies. The result is an excessively litigious system combined with a decision-making process where "choices about environmental protection priorities and goals are buried in thousands of highly technical standard-setting decisions made by agencies and reviewed by courts."[11] This observation reinforces the lesson of the economics of federalism: there should always be a presumption in favor of local solutions to local externalities. In the final analysis, purely local externalities are local problems and can best be dealt with locally.

Minimal Federal Regulation of Interstate Externalities

As discussed earlier, one of the most convincing arguments for federal environmental regulation is the control of interstate externalities. If the external costs are imposed across political boundaries, then the issue should be addressed by a higher level of government. But the presence of interstate externalities does not imply that they

---

10. Aranson, "Pollution Control," p. 384. See also Stewart, "Pyramids of Sacrifice?" pp. 1,210–11.

11. Stewart, "Controlling Environmental Risks through Economic Incentives," p. 158. Stewart has also stated:

> A combination of bureaucratic hearings and review by unelected judges
> is an unlikely process for selecting and implementing measures in the
> general interest. Courts and agencies are buried in lengthy adversary
> hearings that often take many years to resolve. Federalism values are
> severely undermined because interest groups can circumvent state and
> local political processes by bringing federal court actions to force local
> officials to carry out national directives. No one bears clear responsibility
> for decisions. The already severe fragmentation of central authority is

must be corrected by federal regulation that completely usurps the role of local initiatives. The excessive pollution is the consequence of poorly defined property rights. This property rights perspective suggests that the basis for cost internalization could be found through a productive, minimal role for the federal government—the assignment of property rights to states, either to clean air (no pollution from neighboring states) or to the right to pollute across state lines. For example, when only a very small number of states are involved, the federal government's intervention could be limited to the assignment and enforcement of property rights among the states. When the number of states involved is too large for effective bargaining among the states, however, or when states evince a proclivity for acting strategically to obtain payoffs from out-of-state interests, a more interventionist role might be justified. The key to our position is that local governments ought to be allowed to make judgments about their own interests, even if those judgments turn out to be misguided, as long as the costs of these decisions are fully internalized by the particular communities served by the local government.

Assignment and Exchange of Property Rights. Assume two neighboring states, A and B, where industrial air pollution from state A lowers the quality of air in state B. There are two possible allocations of property rights. First, if state A were assigned the right to pollute, state B could still obtain an improvement in its environmental air quality by paying state A to enact and enforce more stringent air quality laws. Citizens in state B would be taxed to pay for their cleaner air. Obviously, this would involve tremendous political battles, but it would force the politicians to assess the actual costs of their actions, a necessary first step for better government. If state B is unwilling to raise the necessary funds for A to agree to stop or reduce pollution across the state boundary, then state A would continue to pollute. The opportunity cost to state A of polluting would be the amount that state B is willing to pay for A to stop. Thus, state A's decision to pollute is not free, and political competition in state A is

---

exacerbated by treating each agency decision as an isolated event to be judicially reviewed on the basis of its separate evidentiary record. The result is a self-contradictory attempt at "central planning through litigation."

Stewart, "Madison's Nightmare," pp. 346–47.

likely to inform constituents of the costs associated with continued pollution.

Second, and more likely in today's political environment, state B might be assigned the legal right to be free of pollution coming from state A. This right could be enforced by either a property rule (injunction) or a liability rule (suit for damages against state A). Either rule would force the internalization of pollution externalities in state A. A property rule would allow state A to negotiate with state B for the right to pollute state B. One can envision state B holding out for progressively higher prices in return for accepting more pollution in the form of lower pollution standards in state A. State A could raise revenues for this right by taxing the polluting industries in state A or, if the state is concerned about adverse consequences on state industries, by use of general fund tax revenues. Taxing the polluting industry on the basis of the pollution emitted would give some polluters the incentive to reduce pollution. Alternatively, state B's right to clean air could be protected by a liability rule under which state A would be forced to compensate state B for damages resulting from excessive pollution from state A. A liability rule raises problems because of the measurement of damages, which in many cases is subjective. Bargaining under a property rule appears to be the preferred allocation of rights, because it requires that all exchanges be mutually beneficial. A liability rule allows for a taking with compensation for objective damages, but not subjective costs.

Although state B might be assigned the legal right to be free from pollution coming from state A, it is certainly possible that pollution sources in state B will also lower the air quality in state B. The use of bargaining to protect or compensate state B for pollution emanating from state A is complicated by the combined effect in state B of pollution sources located in both states. For example, when a liability rule to protect state B is assigned, Coasian bargaining will not be possible until there is some objective way to separate and measure the pollution costs imposed from state A from the pollution costs generated in state B. (The term *Coasian* is derived from Ronald H. Coase, "The Problem of Social Cost.") The federal government or federal courts could be called on to determine responsibility.

Although this framework predicts that bargaining will result in the allocation of resources to their most efficient use—the optimal level of pollution will occur—regardless of the initial allocation of property rights, an important normative policy issue concerns the initial allocation of property rights. Transaction costs can be saved if the initial allocation is to the party or state with the highest valuation of the resources, but this determination is a difficult one. Moreover, basic conceptions of private property suggest that the initial allocation of rights should include the right to exclude others from using one's resources. Thus, this analysis supports a rule requiring the polluting state to negotiate permission from the recipient state before allowing pollution to cross its border into the neighboring state.[12] Such a property rights allocation would probably be popular in today's political environment, but the determination of the politically feasible rule would largely depend on preexisting pollution patterns across state boundaries. Alternatively, a reasonable initial bargaining position would be that the recipient state could force the pollution-exporting state to reduce its pollution to the level that the polluting state would produce if it had the same pollution standards as the recipient states. That is, the recipient state would not be able to hold its neighboring states to a standard higher than it holds polluters in its own back yard.

The potential advantages of such a Coasian system in forcing the internalization of pollution costs across a small number of jurisdictions are intriguing. Of course, such a system would have problems. Bargaining costs might be astronomical, because of political grandstanding. After all, although the property rights would be assigned, the individuals in charge of enforcing them are politicians who do not personally own the transferable property rights. Political competition could force politicians representing states to the bargaining table.

A potential objection to such a Coasian scheme might be the inability of poor states to purchase the right to pollute in rich neighboring states. A clean environment typically becomes more important after basic necessities are met. A related objection might be that rich

---

12. In Georgia v. Tennessee Copper Company, 237 U.S. 474, the Supreme Court prohibited *any* damaging emissions. Such a rule would be much more strict than current environmental law, and it is not obvious that the litigation resulting from such a rule would be more efficient than the current regulatory scheme. The selection of the initial allocation of rights would be an important factor in determining the success of a property rights solution.

states would be able to continue to pollute as they pay poor states to accept their pollution. Similarly, bargaining problems faced by poor states are exacerbated by allocating all pollution control to the federal government, where larger and wealthier states are likely to have greater influence over policies.

Richard Stewart has also considered the possibility of bargaining among states as a solution to interstate externalities (spillovers). Stewart rejects the use of bargaining on the following grounds:

> Bargaining among the states to minimize the losses occasioned by such spillovers is costly (particularly given the complexity and wide dispersion of many forms of environmental degradation), and may do little to improve the lot of states in a weak position (such as those in a downwind or downstream position). These states are likely to favor federal intervention to eliminate the more damaging forms of spillovers.[13]

The problem with this analysis is that Stewart fails to take the crucial first step, *the assignment of property rights*, probably to the "weak" states (hence, making them "strong") and necessarily to the federal government (a limited form of federal intervention).

Small, weak, poor states are better off with the ability to trade pollution rights than they would be without such a right. Without the right to sell pollution rights, poor states would be worse off, because they would have "too little" pollution; that is, these states would be willing to accept a bit more pollution if, in turn, they could also get additional money. Depriving poor states of the ability to make these arrangements makes the residents of such states even worse off. And, of course, if the residents of poor states had a strong preference for high air quality, they could obtain such high air quality by electing officials who imposed tough local standards and refused to sell pollution rights to out-of-state polluters.

Regional Effects and Regional Responses. Frequently, several states have common environmental interests because they are part of the same regional environmental system, such as the Chesapeake Bay region and other watersheds. This is a classic commons problem

---

13. Stewart, "Pyramids of Sacrifice?" p. 1,216.

in that the failure to define property rights means that each state's policies affect the common resource and each state is hesitant to act independently.[14] The federal government can play an important role as a catalyst for regional agreements,[15] and it can also safeguard against certain regions forming alliances against other regions.

The assignment of enforceable property rights is also a potential solution to regional problems involving even a fairly large number of states. It is often assumed that when the pollution from a source in one state imposes costs on numerous other states, the assignment of property rights and the reliance on bargaining could not provide a practical solution to the externality problem. Transaction costs may be too high for meaningful bargaining among the states. This inability to reach a contractual solution, coupled with the usual presumption in favor of the internalization of externalities, means that a response by the federal government may be justified. In contrast with these traditional assumptions about the limitations of contractual solutions, the federal government's role could be limited to the assignment of property rights and the facilitation of bargaining.

Experimental tests of the Coasian theorem with large bargaining groups support such a limited federal role in solving regional environmental problems. A study conducted by Elizabeth Hoffman and Matthew Spitzer to reflect choices made on pollution levels in an externality problem demonstrated that the size of the bargaining group is less of a concern than has traditionally been thought.[16] The results indicated that 93 percent of the bargains among large groups were efficient and that no significant reduction occurred as the group got larger. In fact, bargaining efficiency may have improved as the group size increased. Such information is an affirmation of the potential of federalism in solving environmental problems. The role of the federal government in regional and even nationwide externality situations may be to provide a forum for large groups and rules by

---

14. For a discussion of the tragedy of the commons, see note 30, above.

15. In fact, the Clean Air Act amendments of 1990 envision just such a role for the federal government. See discussion below.

16. See Elizabeth Hoffman and Matthew Spitzer, "Experimental Tests of the Coase Theorem with Large Bargaining Groups," *Journal of Legal Studies* 15 (1986): 149, 151–52. Elizabeth Hoffman and Matthew Spitzer, "Experimental Law and Economics: An Introduction," *Columbia Law Review* 85 (1985): 991.

which to bargain. The Coasian assumptions of enforceable contracts and assignment of property rights must also be a function of this limited federal intervention.[17]

Nationwide Externalities and Federal Regulation

If bargaining solutions to interstate externalities do not appear workable, then the relevant policy issue turns to the precise nature of the federal regulation to be enacted. Such regulations can take a variety of forms, including: (1) centralized command-and-control regulation; (2) federally mandated pollution-based standards for environmental quality in states, where states are free to design their own regulatory apparatus; (3) federally mandated minimum standards for emissions with market-based incentives, where states play little if any role in implementation; (4) a system of Pigovian taxes (the taxing of activities that create negative externalities), imposed by either the federal government or state governments; or (5) some combination of these and other strategies.

Although determining the optimal federal policy when federal regulation is appropriate is beyond the scope of this study, there can be little doubt that federal policy would be better informed if it could draw on the divergent experiences of the states in dealing with other environmental problems. In this regard, governmental intervention on behalf of environmental protection can be viewed as a search for a policy that will produce the optimal amount of pollution. Common sense and economic theory both suggest that the most appropriate source of environmental regulation is not necessarily the federal government but rather the governmental unit most conterminous with the area subjected to the externalities. The economic model of federalism not only provides a way to analyze existing laws; it is also prescriptive, in that it suggests that local laws should satisfy certain conditions. Obviously, all externalities are not national in scope. Thus,

---

17. It has been suggested that a system of resource federations that allows the free transfer of property rights and enforceability of contracts among individuals would be the most efficient solution to environmental problems. See James L. Huffman, "A North American Water Marketing Federation," in Terry L. Anderson, ed., *Continental Water Marketing* (San Francisco: Pacific Research Institute for Public Policy, 1994), p. 145.

the idea of leaving some local control over local externalities seems logical. In fact, the economics of federalism provides strong theoretical arguments for allowing competition among state environmental regulators—the competition may be a source of future wisdom.

5
Restructuring Environmental Regulation

The analysis presented in the preceding chapter suggests that determining the efficient division of regulatory authority within a federal system is not very complicated. In general, regulatory authority should go to the political jurisdiction that comes closest to matching the geographic area affected by a particular externality. Traditional federalist theory tells us that local government regulation should be preferred whenever appropriate, so that regulations reflect the environmental-quality preferences of the affected parties, as well as allow for jurisdictional competition and diversity. Thus, primarily local externalities should be regulated by local governments.

Interstate externalities make up the only area where federal regulation may be superior to local regulation. Interstate externalities often arise when the optimal level of pollution in an upwind or upstream state is greater than the level of pollution tolerated by a neighboring state. Thus, when externalities include the physical imposition of significant costs in states other than the state where the pollution originated, then properly formulated and implemented federal intervention would seem appropriate. The federal response should be limited to the minimum involvement necessary to address the problem. For example, when a small number of states is involved, the federal government's role might be limited to the establishment and enforcement of property rights so that states can bargain and litigate among themselves over the optimal amount of pollution. When numerous states are harmed by or produce multistate externalities, regional or multistate regulatory responses might be more efficient than complete federal domination of the regulatory arena, although the threat of federal regulation may be necessary to force the states to cooperate. In the more limited areas where federal

48

regulation is justified as a last resort, federal regulators should draw on the state-specific expertise of state regulators whenever possible. Finally, federal regulation when adopted as a last resort should attempt to decentralize as much as possible, perhaps through the use of pollution-based statutes and market incentives.

In this chapter, the three major types of pollution—air, water, and land—are evaluated in terms of size of the effect of externalities. The goal is to determine the level of government that most closely matches the size of the externality. This process is then used to evaluate the current allocation of environmental regulatory authority. This model is not an attempt to explain the current environmental regulatory structure.[1] Moreover, it is not an attempt to identify the best environmental policy. The purpose is to suggest ways to reallocate environmental regulatory authority so that better environmental policy, whatever that might be, will emerge from the political process.

Air Pollution

Air pollution has numerous causes and numerous effects. It is often difficult to identify the polluter. Even when the polluter's identity can be determined it is not always easy to identify the extent of the harm. It is nonetheless apparent, however, that some types of air pollution are local and some are interstate.

Smog and Local Air Pollution. Local air pollution and smog can exacerbate respiratory health problems and other public health concerns and can diminish the aesthetic value of the air.[2] The public health and aesthetic costs of smog vary from area to area depending on geographic conditions, demographics, and local preferences. That is, smog and local air pollution are local problems, and substantial benefits can be achieved through the reduction of local air pollution. But as noted earlier, identical reductions in local air pollution are likely to be valued differently in different areas. Accordingly, this would seem to be an ideal situation for allowing local and state

---

1. For such an attempt, see Elliott, Ackerman, and Millian, "Statutory Evolution."

2. See Kenneth Chilton and Anne Scholtz, "A Primer on Smog Control," *Regulation: Cato Review of Business and Government* (Winter 1990): 31.

governments to determine the amount of smog and local air pollution they will tolerate.

Most of the solutions to local air pollution and smog can be implemented with local or state regulations. Professors Kenneth Chilton and Anne Sholtz have identified the six most productive ways to reduce smog: "(1) applying reasonably available control technology to point and area sources; (2) requiring an enhanced inspection and maintenance program for vehicles; (3) instituting transportation control measures; (4) reducing fuel volatility; (5) requiring service stations to install stage II fuel recovery systems; and (6) mandating onboard fuel recovery systems."[3] It is feasible to tailor all these measures, except for mandating onboard fuel recovery systems, to the demands of local citizens, yet the entire history of federal clean air legislation has been to override state and local interests. Implementation of any of these methods at nationwide, uniform levels of intensity means that local areas with little or no smog bear the same costs for cleaning up the environment as polluted areas, but they do not receive the same benefits, because their air is already clean. Alternatively, it could be argued that citizens in some relatively polluted areas do not receive the same benefits because, in fact, they do not value clean air as much as citizens in other areas.

The Clean Air Act amendments of 1990 recognize that different areas have different levels of smog and other forms of local air pollution, as evidenced by the classification system for ozone nonattainment areas.[4] But the classification system still does not recognize that different communities may place different values on clean air. In fact, the Clean Air Act amendments severely constrain the permitted policy responses once an area is placed in a given classification. For example, if an area is classified in a state implementation plan (SIP) as a "moderate ozone nonattainment area," then the SIP must provide for installation of a stage II vehicle refueling system to recover vapor emissions from the fueling of motor vehicles. Thus, reliance on centralized decrees and standards effectively limits the individual state's ability to fashion its own innovative techniques to combat smog and local air pollution, while completely ignoring the

---

3. Ibid.

4. See Title I.

context of the state's fiscal and political situation. The imposition of uniform national standards must reduce the social welfare of many communities.

Recently, state regulations that are more stringent than the federal regulations have been passed. This indicates that there are substantial differences in preferences for clean air across states even with a minimal federal standard, and thus provides support for the federalism policy advocated in this volume.[5] The more stringent state regulations, however, are an example of the cost-externalization problem mentioned earlier. Our analysis suggests that more stringent state regulations should be evaluated in terms of whether the same pollution reduction goal could be achieved in a manner that imposes fewer costs on out-of-state manufacturers. For example, the goals of the stricter California vehicle equipment requirements could be achieved by several alternative policies. Equipment requirements could be replaced by a higher state tax on gasoline, or a requirement that older cars, which produce much more pollution, pay a much higher annual state registration fee. This would have the benefit of making Californians more aware of the costs of controlling pollution in California. Moreover, these taxes and fees could be adjusted to reflect differences in pollution in different parts of the state.

Interstate Air Pollution. The prevention or reduction of interstate externalities is one of the primary rationales for federal regulation, yet few provisions of the Clean Air Act are aimed primarily at interstate externalities.[6] A federalist model of environmental policy suggests that physical externalities across state lines could be dealt with through regional compacts between the affected states. The Clean Air Act amendments of 1990 established a northeastern transport region consisting of eleven states and the District of Columbia, running from northern New England into parts of Virginia. This makes the northeastern part of the United States a single moderate nonattainment area and requires the implementation of reasonable avail-

---

5. For an argument that these more stringent regulations should not be used as evidence to counter the race-to-the-bottom story, see Revesz, "Rehabilitating Interstate Competition," pp. 1,227–33.

6. See ibid., pp. 1,224–27.

able control technology (RACT).[7] This appears to be a regional solution to a regional problem. But as with the classification system used to help diminish local air pollution and smog, the 1990 amendments concerning interstate air pollution do not give the participant states enough flexibility in designing policy responses.

Interestingly, this provision hints at the possibility of a property rights approach to interstate pollution. It requires that the EPA promulgate criteria for measuring the contribution of sources in one area to ozone concentration in other areas. A first step in the use of property rights to control interstate pollution is the assignment of rights. A second necessary step is the ability to identify the wrongdoer. The EPA is charged with developing this technology. If states can be shown to be exporters of pollution, why not allow neighboring states to sue them? The exporters would then be forced either to pay damages or to reduce the amount of pollution exported. Our federalism model suggests that states should be free to choose whatever policy goals they want and the regulatory methods through which the goals are pursued. Given the states' experiences with federal command-and-control regulations, it is unlikely that many states would model their programs after the federal programs.

Acid Rain. Acid rain is caused primarily by factories and power plants that release sulfur dioxide into the atmosphere, which then returns to the earth with precipitation. Because sulfur dioxide is carried through the atmosphere, acid rain knows no political boundaries. Acid rain is an interstate and international phenomenon, but there are serious questions about whether it warrants a regulatory response.

The *potential* for acid rain to cause tremendous environmental harm is not disputed. For example, it is widely recognized that acid rain has caused serious damage in Eastern Europe to lakes, rivers, forests, and even buildings. The extent of the harm in the United States and Canada, while significantly lower than that in Eastern Europe, has been exaggerated by the political rhetoric of environmentalists and by the grandstanding of politicians.

In 1980, the EPA claimed that acid rain had increased the average acidity of northeastern lakes a hundredfold over the preced-

---

7. Section 1.05[2][i].

ing forty years. The EPA alleged that factories and coal-fired power plants located in the Midwest were responsible. The EPA's claim became a catalyst for congressional funding of a ten-year scientific study entitled the National Acid Precipitation Assessment Project (NAPAP).[8] The study found that 90 percent of lakes with high levels of acidity are naturally acidic. The average lake acidity has not increased since the Industrial Revolution—some lakes have become more acidic, while others have become less so. Interestingly, most of the critically acidic lakes in the United States are found in Florida, whose rain is among the least acidic in the eastern United States. The scientific evidence showed virtually no damage from acid rain to crops and forests. The NAPAP report did credit the 1977 Clean Air Act with reducing sulfur dioxide emissions, but it questioned the value of further emission reductions.

In 1990, Congress chose to ignore the NAPAP conclusions. The Clean Air Act amendments of 1990 mandated that power plants reduce their sulfur emissions by 50 percent by the end of the century. Estimates of the annual costs for complying with these regulations range from $5 billion to $8 billion per year. That is a lot of money to solve a trivial, possibly nonexistent problem—especially so considering that the NAPAP study suggested that all the acidic lakes in the Northeast can be limed for about $500,000 per year. Although the 1990 amendments provide for the trading of emission allowances, this is merely a step in the direction of minimizing the costs of achieving the level of emissions mandated by the 1990 amendments, and in no way indicates that the mandated level of emissions is the optimal level.

Such a wasteful policy response would be less likely to occur in a system where states are given more control over their local environmental policies. The problems addressed by the 1990 act are local in terms of where the alleged harm is taking place, yet the act imposes costs throughout the national economy. The political dynamics of centralized environmental regulation, dispersed costs, localized benefits, and environmentalists' hyperbole all combined to generate a wasteful policy.

---

8. For a description of the project and a summary of its major findings, see J. Laurence Kulp, "Acid Rain: Causes, Effects, and Control," *Regulation: Cato Review of Business and Government* (Winter 1990): 41.

Of course, the fact that acid rain creates a physical interstate externality suggests that this may be an ideal situation for a federal response. The localization of the harm, however, and its easily quantifiable scope in terms of the costs of correction indicate that the federal response should be limited. For example, federal law could simply provide for lawsuits by states against the polluters for the costs of correcting the problem. Liability could be prorated on the basis of contribution to the pollution. This would protect the northeastern states from acid rain caused by pollution and would provide the proper incentives for polluters to engage in the optimal amount of pollution avoidance. It should be noted that a liability rule should be preferred to a property rule because of the enormous difference between the cost of correcting the harm (such as liming lakes) and the cost of avoiding it (installing scrubbers). Granting the northeastern states a property rule, enforced by injunction, would allow them to hold up the Midwest sources of sulfur dioxide.

In conclusion, the federal domination of air pollution regulation was strengthened by the passage of the Clean Air Act amendments of 1990. Most of the regulations continue to reflect the failed command-and-control form of centralized regulation. The estimated costs of compliance with the new regulations are staggering, and there is little reason to believe that there will be a significant return on the resources invested in the process.

Water Pollution

Water pollution is difficult to regulate for many of the same reasons as air pollution. Not only do waterways flow between states, but because all water eventually makes its way to the ocean, water pollution in any type of lake or stream is potentially an international problem. Yet, as is the case for air pollution, there are numerous ways to control water pollution at the state and local levels without federal domination of the field.

Localized Water Pollution. Consistent with our model, pollution with purely local effects should be dealt with locally. In particular, the environmental problems involving wetlands and sewage disposal are generally local. There is no reason to believe that local authorities do not have the appropriate incentives or capability to deal with these problems. Moreover, for a variety of reasons, the discharge of pollutants in waterways may be controlled more effectively at the state and local levels.

For at least two reasons, it is easier to establish locally based, market-driven incentives for dealing with water pollution than for air pollution. First, as noted above, a perennial problem with regulating air quality lies in allocating the costs imposed on various localities to particular sources. Allocation presents problems for regulators both because it is difficult to identify the various sources of environmental harm and because it is difficult to allocate damages among these various sources. It is often easier to identify and trace water polluters than air polluters, however, and it is easier to quantify the damage being done to a particular water source.

Many economists have suggested, for example, that marketable permits be used as devices for improving the efficiency of regulation.[9] It is easy to conceptualize a regulatory framework in which marketable permits are used to implement environmental policy. First, regulators must identify a target level of environmental quality and translate that goal into policy by establishing a goal for total allowable emissions.[10] Permits giving owners the right to specified levels of pollution are then allocated to individual firms. Firms are allowed to trade these permits. Where the market for these permits is allowed to work, the total cost of achieving a particular environmental goal will be minimized by a permit policy.[11]

Because it is relatively easy to identify firms that are discharging pollutants into a particular water source after the target level of environmental quality is determined, it would be relatively easy to establish how many permits could be issued. Because of the difficulties in identifying all the sources of air pollution, by contrast, it

---

9. See Robert Hahn, "Designing Markets in Transferable Property Rights: A Practitioner's Guide," in Joeres and David, eds., *Buying a Better Environment: Cost Effectiveness Regulation through Permit Trading* (Madison, Wis.: University of Wisconsin Press, 1983): 83–97.

10. Robert Hahn, "Economic Prescriptions for Environmental Problems: How the Patient Followed the Doctor's Orders," *Journal of Economic Perspectives* 3 (1989): 95, 96.

11. Michael Montgomery, "Markets in Licenses and Efficient Pollution Control Programs," *Journal of Economic Theory* 5 (1972): 395.

might be difficult to determine how many permits should be issued and who should receive them.

A localized alternative to the command-and-control regulation contained in the clean water act would encourage local regulators to issue tradable permits. Tradable permits would allow firms to benefit from being able to reduce pollution at reduced cost. Those firms that could reduce their emissions at low cost could sell their permits to firms that cannot reduce their emissions as efficiently. Similarly, tradable permits allow organizations and individuals with very low tolerances for pollution to transform their own preferences into reality by buying up permits from polluters and refusing to resell them.[12] Finally, tradable permits give firms incentives to develop new technologies for controlling pollution, both to avoid having to buy permits and to profit from the sale of permits already owned. The tradable sulfur dioxide emission allowances under the acid rain title of the clean air act are designed to achieve these same benefits.

Some evidence suggests that tradable permit programs should work for water pollution as well. The state of Wisconsin organized a permit system on the Lower Fox River in 1981. The Lower Fox River, which runs from Lake Winnebago to Green Bay, is lined with ten pulp and paper mills and four municipalities that discharge significant pollutants into the water. Early studies showed that savings on the order of $7 million per year would result from the permit program. Unfortunately, regulatory restrictions on the permit program, particularly on the marketability of the permits, have limited the effectiveness of Wisconsin's experiment. For example, trading is limited by location, and regulations restrict the number of firms allowed to trade for rights to pollute at points along the river where demand for permits would be highest. This creates an extremely thin market, which diminishes the value of the permits.[13]

Similarly, permits are good only for five years, while transfers must be for at least one year, and it is not clear how renewals will be made, or how trading will affect renewal rights.[14] This uncertainty understandably impedes the market. Moreover, firms wishing to ac-

---

12. Ibid.

13. Hahn, "Economic Prescriptions," pp. 97–98.

14. Ibid.

quire permits must justify to regulators the need for new permits, and exchanges are not permitted for firms that want only to reduce their operating costs. These policies run counter to the basic theory of marketable permits. In light of all the regulatory impediments to free trading in these permits, trading has been virtually nonexistent. In the first six years of the program, there was only one exchange in permits.

Nonetheless, it seems clear that permit-exchange programs such as the one tried in Wisconsin could work if the permits were freely tradable and implemented in a rational manner. Such permits allow individual firms the flexibility they need to adjust their emission levels to environmental concerns in the most efficient way possible. This is just one of the many innovative responses that states would be free to experiment with in designing their optimal environmental policy.

Interstate Water Pollution. Obviously, no individual state can regulate a major river like the Mississippi. Instead, a consortium of states must act together to set environmental standards. In fact, "most major lake and river systems are the subject of intensive water quantity and quality managements under watershed systems established under state law or interstate compacts."[15] The role of the federal government is to act as an arbiter among the states to ensure that a level playing field is created. For example, if a region chooses to adopt a tradable permit program for the reduction of water pollution, the role of the federal government would be limited to supporting the market for permits. Thus, consistent with the regulatory model of environmental federalism presented in this volume, we suggest that the federal government abandon its role as the developer of centralized command-and-control environmental regulations and instead assume a role as a facilitator of bargaining among states.

In particular, attention must be given to the fact that some states have locational advantages over others when it comes to polluting. States lucky enough to be located upstream from other states have incentives to increase their discharge levels, since the costs associated with such discharges are borne by downstream states. It should be the role of the federal government or the federal courts to establish

15. Ackerman and Stewart, "Reforming Environmental Law," p. 187.

property rights in clean water, so that downstream states could assert claims against their upstream neighbors.

The most effective way for the federal government to discharge its responsibility to facilitate the operation of the federal system would be to assign ownership rights in water to individual states. In this way states through which polluted water passed could assert a cause of action against the states responsible for the pollution. Downstream states would be able to sue upstream states for the costs associated with the pollution being sent downstream. In turn, the states that serve as forums for polluters would have several options— including state command-and-control regulation, permit systems, or use of tax revenues to obtain funding from polluters in the form of payments for the permits described above. Imaginative and innovative applications of this property-rights system could work for most types of water pollution.[16] Our federalism model assumes that states will have the incentives to pursue such solutions once they are given the authority and responsibility for environmental quality in their jurisdictions.

Land Pollution

The control of land pollution through the regulation of the disposal of solid wastes has traditionally been a state or local issue. But as in air and water pollution, the federal government's regulation of land pollution has increased. Much of this regulation has been adopted at slightly later time periods than the federal regulation of air and water pollution. The later response, however, has been coupled with unprecedented federal involvement in local land use decisions and draconian liability provisions. This section assesses the current allocation of regulatory authority from the perspective of our federalism model.

---

16. An innovative and perhaps more efficient solution may be a system of regional federations. Such federations would be immune to state and national boundaries. Their responsibility would be the facilitation of a property transfer market over one resource (for example, regional water market federations). Placing property rights into the hands of individuals leads to a market-based solution and eliminates the role of both state and federal governments. This is federalism at its maximum potential. See James L. Huffman, "A North American Water Marketing Federation," in *Continental Water Marketing*, Terry L. Anderson, ed., 1994, p. 145.

Solid waste landfills and toxic waste dumps are the most common land pollution sources. One important consequence of land pollution is the possibility of groundwater contamination—a serious problem, since groundwater serves as the community's source of drinking water and provides in effect an underground river for transporting harmful substances away from their sources to other areas. Because land pollution and the potential for groundwater contamination are very localized phenomena, our federalism model leads to the argument that these externalities should be regulated exclusively by state and local jurisdictions.

Garbage and Landfills. Under most circumstances, landfills are the most local of pollution externalities and thus should be regulated by local and state governments. The decision about whether to accept garbage from outside the locality or state should be determined by the smallest possible political jurisdiction. If an area wishes to specialize in landfills, or wishes to have mandatory recycling, regardless of whether it is cost efficient, our federalism model argues that the area should be allowed the freedom to implement its own policy. The more flexibility afforded local governments in dealing with garbage, the cheaper and safer the local environment. In fact, there is considerable variety across states in the standards imposed on landfill operators, as indicated by the response to the recently promulgated EPA rules governing landfills: "Impact will be biggest in those states, mostly in the South and West, without tough laws. Those with strict laws 'will hardly' notice," says Ed Repa of the National Solid Wastes Management Association.[17] Clearly, the relative lack of an interstate externality suggests that there should be no role for the federal government in regulating landfills.

The biggest environmental threat posed by landfills is the contamination of groundwater. Although aquifers can be enormous, the dispersion of pollutants within them is usually slow and confined to small areas. Thus, groundwater contamination from landfills should in most cases be handled by state and local regulation of landfills. Where an aquifer extends into more than one state, each state should be given the right for the aquifer to be free from landfill contaminants.

---

17. Rae Tyson, "EPA Signals 'End of the Town Dump,'" *USA Today*, September 11, 1991, p. 1.

A litigant could enforce these regulations either by asking the court to order an injunction against the wrongdoers or by suing the polluters for damages.

Given the proper incentives, landfill operators will employ modern solid waste management techniques to reduce the likelihood of groundwater contamination to appropriate levels. These incentives can be found in privatization and regulation under the common law of torts. Holding private landfill owners liable for groundwater contamination is a powerful incentive, and it would serve as a more efficient means than federal regulation of ensuring a safe supply of groundwater.[18]

A more difficult issue concerns how to deal with state statutes like the Maine provision outlawing juice boxes that are alleged to be nonbiodegradable. On the one hand, states should have some latitude to control the influx of pollutants by out-of-state manufacturers. On the other hand, there is always a danger that states will use their police power to engage in cost externalization in favor of local industries over out-of-state producers. Thus, it is not possible to formulate a universal rule to govern situations like this. Instead, federal courts must balance the local interests against the general principle favoring unfettered interstate commerce, always keeping in mind that states are susceptible to political pressures that may cause them to regulate out-of-state producers in order to transfer wealth to local rivals. Thus, courts should require that any regulation that has extraterritorial effect use the least restrictive means available to accomplish its objectives. Similarly, courts should inquire whether a state's regulatory regime has a disproportionate effect on out-of-state producers. And finally, courts should inquire whether the legislation is part of a

---

18. This incentive is absent for public landfills, which are exempt from liability. It is also interesting to note that private landfill operators were disappointed that the recently announced EPA rules for landfills were not as strict as they had hoped:

> Major trash companies, which have been upgrading their dumps in anticipation of stiffer rules, contend that lower-quality dumps owned by municipalities could stay open for some time.
>
> That will delay the lucrative shift of some trash to the private companies' dumps. Still, the industry hopes that the rules will accelerate closings, boosting profits of big dumps owned by [private companies].

Rose Gutfeld and Jeff Bailey, "EPA Sets Rules for Pollution Curbs on State Landfills," *Wall Street Journal*, September 12, 1991, p. A8.

larger, internally consistent regulatory framework aimed at environmental policy or whether it appears to be an ad hoc measure aimed at accomplishing some other end.

Toxic Waste Sites. Because toxic waste sites are essentially specialized landfills, the same type of local control should prevail here as over general purpose landfills. Imposing a complex, centralized federal regulatory system on thousands of highly localized sources of toxic waste, underground storage tanks, and pesticides and herbicides makes no sense. We support the idea of federal tracking of the interstate movement of pollutants, but we believe such tracking should be the only area of federal involvement in dealing with hazardous waste sites. Several aspects of the governmental response to hazardous waste could be improved by shifting regulatory authority to state governments.

First, federalism would allow for a more flexible response to the toxic waste problem. At present, the EPA has the authority to determine what constitutes a fully cleaned up Superfund site. This is a mistake. Individual states should be given this authority, as some areas may have a higher tolerance for impurities than others. Local residents may be willing to accept tax abatements in exchange for living near partially cleaned or contained sites. A federalist approach would not only reduce costs for industry; it would also direct cleanup dollars where they are most wanted.

Second, analysis of the structure of risks addressed by the Superfund reveals numerous opportunities for local governments to make the necessary policy trade-off of expensive cleanup versus simply fencing off a site and prohibiting access. In a detailed study of risks addressed by the Superfund, James T. Hamilton and W. Kip Viscusi found that the EPA's risk assessment is based on unnecessary assumptions about future use of the site:

> Most of the political pressures that generated the impetus for the Superfund program arose because of the concern of existing populations for the risks they believe these sites currently pose. Consideration of the risk assessment for Superfund sites indicates, however, that it is not the existing risks that are most salient. Rather the dominant risks arise from future risk scenarios that generally involve alternative uses of land. Indeed, these future risks account

for 90 percent of all the risk-weighted pathways for the Superfund sites in our sample. Chief among these future risks is that there may be future residents on-site. The underlying assumption driving the EPA risk analyses is that there will be new residential areas on existing future Superfund sites, where there are not currently such residential areas.[19]

Obviously, lest builders consider erecting homes on such a site, the potential risks to residents could be avoided by local government condemnation of the property.

Third, individual states are far more likely to be concerned about the effects of cleanups on local businesses. As discussed earlier, zero pollution is an invalid goal, and striking the appropriate balance between business and resources is best done at the local level. For example, Superfund imposes retroactive liability on firms that have caused hazardous waste. Firms are now being held responsible for damage done thirty or forty years ago, for engaging in activities not only perfectly legal at the time but often sponsored or encouraged by the government. The federal government is, in effect, abrogating long-standing agreements between states and companies, in which the states agreed to accept certain levels of toxic wastes in exchange for economic growth.

Finally, unlike the EPA, state officials actually live where toxic waste sites are located. As such, these officials have a greater incentive to see that funds allocated to cleanups are used for cleanups and not wasted in costly and unproductive litigation. The results of Superfund are truly deplorable. A program that was to cost $5 billion and to last for five years is now expected to cost $1 trillion and take at least fifty years to complete. Even more shocking, of more than 1,200 sites on the National Priorities List, only 33 have been fully cleaned up. Now EPA estimates that as many as 10,000 additional sites may have to be added to the list. It is not clear that these sites are serious health threats. Given the EPA's dismal record, however, it is clear that there is no way all the sites will be cleaned up. Once again, allowing states to take responsibility for cleaning up toxic

---

19. James T. Hamilton and W. Kip Viscusi, "Human Health Risk Assessments for Superfund," Conference Paper, AEI Conference on Reforming Superfund, American Enterprise Institute, June 3, 1994.

waste sites would ensure that the most pressing problems are addressed first. The EPA, conversely, seems incapable of prioritizing sites in any rational way.

Toxic waste is a local problem that should be treated locally. Individual states should not only be allowed to determine how scarce cleanup resources are allocated within their states: they should also determine whether they will be net importers or exporters of such waste. Some states might be willing to accept higher levels of toxic waste in the form of looser environmental standards, in exchange for new industry, or in exchange for dollars. Such flexibility is impossible under the current centralized regulatory framework imposed by Comprehensive Environmental Response, Compensation Act of 1980 (CERCLA).

This analysis suggests that the entire Resource Conservation and Recovery Act (RCRA) and CERCLA systems should be dismantled. The current system imposes tremendous costs with little return. The one provision that should be retained is the manifest system that could help states maintain the integrity of their political boundaries with regard to the hauling of trash and toxic waste.

6
Conclusion

The environmental policies that we actually observe are at odds with the theory of federalism articulated in this volume. One of the most important attributes of a properly functioning federal system is that local governments are given autonomy to tailor regulatory solutions to local problems and concerns, leaving the federal government free to address multistate problems.

There are several reasons why local governments should be permitted to address environmental issues that have primarily localized effects. First, different localities are likely to have different preferences and concerns. Decentralized government through a federalist system is far more responsive to local needs and concerns. For example, some communities might prefer to trade off environmental quality for more employment or greater revenue. Local control over environmental issues would permit this. Second, local control is beneficial because state and local governments will engage in healthy competition along a number of vectors. They will compete to attract new business, they will compete for jobs and revenues, and they will compete to offer residents better environmental quality. By contrast, the centralized, monopolistic command-and-control apparatus of the federal government does not offer citizens the benefits of competition. Finally, where local decision-making authority is replaced by federal regulation, rational local officials will compete to obtain wealth transfers from other localities. Every locality will consume resources in lobbying for environmental policies that produce local benefits, regardless of the consequences for the nation as a whole.

Although this perspective may seem radical to some environmentalists, it has a long history in American constitutional law and theory. As James Wilson explained at the Pennsylvania ratifying convention, "Whatever object of government is confined in its operation and effect, *within the bounds of a particular State*, should be considered as belonging to the government of that State."[1] As Professor McConnell has observed, Wilson's view, which was the dominant perspective in the debates of the period, "stands in marked contrast to the modern tendency to resolve all doubts in favor of federal control."[2] And as we have shown, many important federalist problems are problems of purely local concern, which should be regulated at the local level. In particular, many of the sites targeted by Superfund are contained within the confines of a single state. The cleanup of these sites presents issues of purely local concern, and the economic theory of federalism articulated in this volume, which is entirely consistent with the framers' design, would confine authority over this issue to local regulators. Under our proposed policy, if the federal government wants to intrude on local decision-making authority concerning the cleanup of local sites, it should confine itself to lending expertise and providing funding. It should refrain from imposing substantive standards or imposing legal liability.

Clearly, however, not all environmental problems should be addressed by local authorities—like policy issues generally. Where one state is producing environmental hazards not contained within its borders, a national response may be called for. In most instances, however, that response should be limited to the assignment of property rights and the facilitation of bargaining. A federal response is appropriate on occasion; but the federal response should be tailored to particularized environmental and federalism concerns.

---

1. Quoted in Michael W. McConnell, "Federalism: Evaluating the Founders' Design," *University of Chicago Law Review* 54 (1987): 1,484, 1,495 (emphasis in original).

2. Ibid.

About the Authors

HENRY N. BUTLER is the Fred and Mary Koch Distinguished Teaching Professor of Law and Economics at the University of Kansas Schools of Business and Law, and he is the director of the schools' Law and Organizational Economics Center. He is the author or co-author of several books and dozens of journal articles concerning corporate governance, economic analysis of law, and the impact of government regulations on business activities.

JONATHAN R. MACEY is the J. DuPratt White Professor of Law at Cornell University and is director of the John M. Olin Program in Law and Economics at Cornell Law School. He specializes in corporate law, banking regulation, law and economics, and political theory and public choice. Mr. Macey is the author of more than ninety scholarly articles, as well as numerous articles in publications such as the *Wall Street Journal* and the *Los Angeles Times*. In 1995 he received the Paul M. Bator Award for Excellence in Teaching, Scholarship, and Public Service from the Federalist Society for Law and Public Policy Studies.

*This book was edited by Cheryl Weissman of the publications
staff of the American Enterprise Institute.
The text was set in Bodoni Book.
Coghill Composition Company of Richmond, Virginia,
set the type, and Edwards Brothers, Inc.,
of Lillington, North Carolina, printed and bound the book,
using permanent acid-free paper.*

AEI Press is the publisher for the American Enterprise Institute for Public
Policy Research, 1150 17th Street, N.W., Washington, D.C. 20036; *Christopher C. DeMuth*, publisher; *Dana Lane*, director; *Ann Petty*, editor; *Leigh Tripoli*, editor; *Cheryl Weissman*, editor; *Lisa Roman*, assistant editor (rights and permission).

www.ingramcontent.com/pod-product-compliance
Lightning Source LLC
Jackson TN
JSHW011942131224
75386JS00041B/1515